SpringerBriefs in Economics

For further volumes:
http://www.springer.com/series/8876

Pasquale Pazienza

The Relationship Between FDI and the Natural Environment

Facts, Evidence and Prospects

Springer

Pasquale Pazienza
Department of Economics
University of Foggia
Foggia
Italy

ISSN 2191-5504 ISSN 2191-5512 (electronic)
ISBN 978-3-319-04300-5 ISBN 978-3-319-04301-2 (eBook)
DOI 10.1007/978-3-319-04301-2
Springer Cham Heidelberg New York Dordrecht London

Library of Congress Control Number: 2013957994

Printed on acid-free paper

Springer is part of Springer Science+Business Media (www.springer.com)

To my beloved children
Thomas George and Emma Victoria,
my wife Claire,
and all my family

Contents

Abbreviations

BITs	Bilateral Investment Agreements
BOD	Biological Oxygen Demand
CEOs	Chief Executive Officers
CO_2	Carbon Dioxide
EKC	Environmental Kuznets' Curve
FDI	Foreign Direct Investment
IF	International Firm
IMF	International Monetary Fund
JV	Joint Ventures
LDCs	Less Developed Countries
M&As	Merger and Acquisitions
MAI	Multilateral Agreement on Investment
MNC	Multinational Corporation
MNCs	Multinational Corporations
NAFTA	North American Free Trade Agreement
NGOs	Non-Governmental Organisations
NIMBY	Not in My Backyard
NOx	Nitrogen Oxides
ODA	Official Development Assistance
OECD	Organisation for the Economic Cooperation and Development
PIF	Private International Finance
SO_2	Sulphur Dioxide
TNC	Transnational Corporation
UNCTAD	United Nations Conference on Trade and Development
USDIA	United States Direct Investment Abroad
WB	World Bank
WIPS	World Investment Prospects Survey
WIR	World Investment Report

Chapter 1
Introduction

One of the most relevant aspects of the globalization phenomenon is represented by the dynamic of financial capital flows around the world. Although the term "international financial flow" refers to a series of forms of capital, in which Official Development Assistance (ODA)[1] and tools of the so-called Private International Finance (PIF), such as portfolio equity investment[2] and debt finance[3] are considered, it often centres on Foreign Direct Investment (FDI).[4] The reason why most of the available literature refers to FDI lies in the fact that, as a form of private capital, it accounts for the greatest part of the PIF to

[1] ODA is represented by flows of official financing conceded with a grant element of at least 25 % and administrated with the main aim of promoting economic development and welfare in developing countries. ODA flows comprise of contributions given by donor government agencies at all levels to developing countries ("bilateral ODA") and to multilateral institutions. ODA receipts comprise of disbursements by bilateral donors and multilateral institutions (OECD 2007).

[2] Portfolio investment refers to the category of international investment that covers investment in equity and debt securities, excluding any such instruments that are classified as direct investment or reserve assets (OECD 2007).

[3] It can be referred to either commercial loan or bonds. The first refers to loan financing to developing countries by commercial banks, export credit agencies, other official institutions in association with other agencies or banks, or the World Bank and other multilateral financial institutions. The latter refers to a debt tool that usually gives the holder the unconditional right to fixed money income or contractually determined variable money income. With the exception of perpetual bonds, it also provides the holder with an unconditional right to a fixed sum as repayment of principal on a specified date or dates. In international finance a bond can typically assume the specific form of structured bonds. These have some characteristics that are designed to attract a certain type of investor and/or take advantage of particular market circumstances.

[4] For now, we simply define FDI as investment made by a company of a country in subsidiary or joint venture firms abroad. A more detailed definition and description of its main features will follow in the next sections of this work.

P. Pazienza, *The Relationship Between FDI and the Natural Environment*,
SpringerBriefs in Economics, DOI: 10.1007/978-3-319-04301-2_1,

emerging economies.[5] Its relevant and beneficial role in countries growth and development processes is generally recognized and reported in terms of job creation, introduction and spread of innovation and new technologies, transfer of intangible resources such as better practices and new methods of organization, which result in production efficiency improvements and increase in competitiveness.

However, we should also consider that FDI is characterized by a "hidden aspect", which is not always taken into proper consideration. According to some of the last available reports of the United Nations Conference on Trade and Development (UNCTAD), it can be appreciated that FDI has always traditionally and significantly relied on the use of natural resources (especially in agriculture, mineral extraction, fuel and chemical production). Although the evolutionary dynamic of the last 20 years or more has shown a structural shift of FDI flow towards the service sector, which is generally believed to be less resource-use intensive, it can be observed that a relevant amount of FDI is still reaching developing countries and especially those sectors primarily based on the use of natural resources. Furthermore, as is provisionally referred for the years ahead, FDI flow is expected to increase in the primary sector, and particularly in the extractive industries of resource-rich countries (UNCTAD 2007, 2004).

In addition, particularly in the last decades increasing and accelerating trends of environmental degradation have also been recorded, which are widely referred to in a number of scientific reports and studies. The global warming generated by greenhouse gas emissions seems to be just the synthesis of a variegated series of environmental problems, which range from deforestation and biodiversity loss to ice melting and the change in sea levels (UNEP 2007). As is generally recognized and claimed, these patterns of environmental degradation are also the result of widespread economic activities worldwide. The recognition that FDI relevantly contributes to this dynamic raises some concerns on the effects associated to the FDI phenomenon and, particularly, on its environmental effects, whose identification is of crucial importance in identifying and implementing appropriate governmental policies. In fact, whether FDI is really functional for development and, in particular, for sustainable development, depends especially on the way it is managed by the receiving countries' governments. In other words, it depends on their vision of economic development and environmental conservation

[5] FDI is the single largest and fastest growing component of private capital flow, especially in developing economies, despite the effect of the global economic downturn of the last few years (UN 2012, 2011). If we consider, for example, the period between the early 1980s and the early 2000s—corresponding to the time span considered in our empirical analysis which will be presented in the next sections—its inflow to Low-Income Countries (LICs) averaged only 0.2 % of their GDP in the early 1980s, but rose to more than 3 % by the end of 2006, showing a more than tenfold increase. The other private flows (including workers' remittances), instead, were more than triple during the same period. In fact, they rose from 1.1 % of the LICs GDP in the early 1980s to 3.6 % in 2006 (Dorsey 2008: 47). In more recent years, between 2009 and 2010, the FDI quota arriving to developing and transition economies reached more that half (53 %) of global FDI flow (UNCTAD 2012, 2011).

management that is on their policy and regulatory frameworks. However, the identification of the environmental implications resulting from FDI movement—and more broadly from the transfer of international capital flows—to other countries and particularly to developing areas is not easy and gives rise to complex arguments and contradictory perceptions and views.

On the one hand, for example, investors—especially those who move resource-seeking FDI—basically find their motivation in searching for those countries which can ensure the highest level of economic returns. As is quite often perceived and sometimes observed, these countries are normally those having a relevant endowment of natural resources and a feeble or ineffective environmental regulatory framework (UNCTAD 2007, 2004). Such a condition potentially represents a high threat for the local communities and the natural environment upon which their lives rely.[6] It is also perceived and observed that countries' economic growth induced by international investment is often accompanied by "industrialized countries' style" or "western-style" consumerism, which can potentially represent a further contribution to worsen the equilibrium of the world's natural system, the earth's climate conditions and the security of food supply as a result.[7]

On the other hand, international investment is also felt to bring benefits to the natural environment. In particular, the FDI movement from developed to developing countries can facilitate the transfer of more modern technologies, which guarantee a greater efficiency in the use of natural resources and energy, together with the minimization of waste and residuals from the production process. If this is the case, developing countries might avoid some of the more damaging phases of the industrialization process well known to those industrialized countries which are still bearing the clean up cost of their natural environmental systems.

However, it seems that the environmental reflection within the FDI issue has generally suffered from a lack of adequate attention and has often been left aside and unconsidered.[8] For example, in analysing the rapidly increasing dynamic of

---

[6] With regard to this, let us think of situations such as, either the implementation of a large construction project which displaces local people from their territory, or those cases, especially observable in the area of the Amazon forest in Brazil and in various other African countries, where indigenous communities defencelessly watch the change and disappearance of their homelands as a result of timber companies' activities.

[7] In this perspective, international investment can also lead to a lifestyle change for local communities, which might begin to express preferences for the consumption of industrial polluting goods such as cars, paper, plastic, etc. As a result of this situation, there is an increase in industrial activity and in turn a growth in pollution emissions.

[8] It is worth highlighting at this point that very often the literature discusses more the environmental effects of globalization, while referring to the relationship between trade—rather than FDI—and the environment. As is generally observed, the existence of a link between FDI and trade exists and is based on their relationship of either complementariness or substitutability. On the one hand, according to the earlier view, the shrink of trade barriers promotes the growth of trade and FDI. On the other hand, evidence of an inverse FDI-trade relationship exists. The debate refers to factor proportion differences, protectionist behaviours, and proximity-concentration trade-off of countries as main aspects influencing this dichotomy. Empirical evidence varies a lot depending on the qualification of sample and proxies used, thus showing huge difficulty in finding

the international capital flow towards Asia since the beginning of the 1990s, and its reversed course following the economic crisis recorded in the region during the second half of that decade, commentators and experts referred to the outgrowth of those countries' financial regulatory structure as a main explanation. Only a few people paid attention to the other critical aspect represented by the relationship between international capital flow and the environment. Their aim was to understand the extent to which the huge amount of international finance flown to the countries of that developing region destabilized the ecological foundations of these emerging economies (i.e. Shahbaz et al. 2011; French 1998).

As will be reported later in the sections analyzing the literature review on the relationship between FDI and the environment, much of the current debate focuses on the conditions characterizing the process of FDI location decision, the resulting competition between countries for FDI and how this effects environmental standards and regulations. Among the various aspects the literature considers within this field of argument, the "pollution havens" hypothesis seems to be the most investigated. Very briefly, this hypothesis states that FDI moves to those countries where more lenient environmental regulations offer the advantage of producing at a lower cost. As a result, countries may compete for FDI, thus getting involved in a "race to the bottom", which is a further hypothesis occurring when countries intentionally undervalue their environmental assets and lower the stringency of their environmental regulations with the aim of bringing in FDI, thus generating an increase in pollution and environmental degradation. As will be referred later in the appropriate section, however, empirical studies have been and are still unable to systematically prove the existence of the various hypotheses, thus the achievement of a universally accepted conclusion is still missing. Among the various difficulties occurring in the identification of the above considered aspects, the use of aggregate data of investment flows, the excessive focus on site-specific environmental impact and the consideration of emissions related to a few industrial pollutants are the most relevant. However, there is plenty of evidence that pollution-intensive industries do have location preferences for low environmental standards and are able to influence governments to create lenient environmental regulations (WWF 2001).

As a result of the difficulty to empirically investigate the very complex and dynamic interaction between increasingly mobile production and environmental regulation, some feeling that the "pollution havens" hypothesis debate has generated a policy stasis, by attempting to pursue indemonstrable evidence, exist.

(Footnote 8 continued)
a generally valid pattern in the FDI-trade link (Michi et al. 2006, 2005). Having said this, we highlight how the issue of trade goes well beyond the scope of our work. For this reason, we do not go any further in its analysis. However, where possible and retained useful, the aspect of trade—in the form of the various proxies it can assume—will be taken into consideration in the next chapters where the empirical tasks are reported. In agreement with other works, in fact, FDI does not occur in a vacuum and the decomposition of its relationship with the environment into scale, structural and technology effects—as they will be explained later in this chapter—would be very difficult without considering the FDI links with other aspects among which trade appears to be one of the most relevant (OECD 2002).

Indeed, it seems that the excessive focus on this hypothesis has driven the discussion on FDI-environment relationship away from what could be perceived as other more relevant questions such as, for example, the identification of more specific linkages between FDI-induced development and the environment, the regulatory capacity and the environmental limits within which economic activities can take place, and the planning of resource use.

Having said this, FDI is still widely and convincingly thought to be beneficial for societies and their natural environments, this being an incentive for negotiators and policy decision-makers to support the pointlessness of setting up environmental restrictions in international investment agreements. However, it must be avoided that FDI-induced economic growth is achieved at the expense of societies and their natural environment. For this reason, a call for better management and governance of FDI should be on the agenda of institutions and scientific research with the aim of identifying methods and ways of making FDI really work for sustainable development.[9]

The objective of this work is to contribute to the understanding of the relationship between FDI and the natural environment by providing a review of the specific literature. To this purpose, the rest of this book is organized as follows. Chapter 2 will be dedicated to representing and explaining the introductory and main aspects of FDI. More specifically, its definition, main features, measures and the socio-economic effects will be overviewed. In the Chap. 3, the environmental effects of FDI will be the subject of attention; with regard to this, technique, scale and composition effects will be analysed in light of continually declining the literature development. The competition for FDI and its effect on environmental standards will be discussed in Chap. 4. Here, the literature related to the "pollution haven", the "race to the bottom" and the "regulatory chill" hypotheses will find its discussion corner. Chapter 5 will be dedicated to the Cross-Border Environmental Performance and, within this framework, the literature on the "pollution halos" (or the "race to the top") theory will be analysed. The geographical and sectoral distribution of FDI will be the subject of analysis in the Chap. 6 where a focus on the trends and prospects of FDI will be developed. Chapter 7 will draw some concluding considerations and will highlight some ideas for a research agenda.

[9] This might sound like a call for a typical "green growth" approach "decoupled" from environmental degradation. Of course, it is not since a response to the growth dilemma based on "decoupling"—that is sustained economic growth with repeatedly diminishing material throughput—appears unrealistic. In fact, some recent studies produce empirical evidence to clearly show how "decoupling" is not happening at all. This is particularly true for the case of "absolute decoupling", this implying that resource impacts decrease in absolute terms as a considered economy grows. However, even "relative decoupling"—occurring when resource impacts still rise but at a slower pace than the GDP—is not observed apart from some specific cases of resource use (e.g. Jackson 2009).

References

Dorsey T (2008) A capital story. IMF Financ Dev 45(2):46–49

French H (1998) Investing in the future: harnessing private capital flows for environmentally sustainable development. Worldwatch paper no. 139, Worldwatch Institute, Washington D.C

Jackson T (2009) Prosperity without growth. Economics for a finite planet. Earthscan, London

Mihci H, Cagatay S, Koska O (2005) The impact of environmental stringency of the foreign direct investments of the OECD countries. J Environ Assess Policy Manage 7(4):679–704

Mihci H, Cagatay S, Koska O (2006) FDI-trade-environment linkages: changing flows between EU(15), enlarged EU and non-European countries. In: Discussion paper for the 8th European treade study group (ETSG) annual conference, Vienna, Austria

OECD (2002) Environmental benefits of foreign direct investment: a literature review. In: Document of the working party on global and structural policy, OECD, Paris, France

OECD (2007) Glossary of statistical terms. OECD, Paris

Shahbaz M, Nasreen S, Afza T (2011) Environmental consequences of economic growth and foreign direct investment: evidence from panel data analysis. MPRA paper no. 32547

UN (2011) World economic situations and prospects 2011. United Nations Publication, New York

UN (2012) World economic situations and prospects 2012. United Nations Publication, New York

UNCTAD (2004) World investment report 2004. In: United Nations conference on trade and development, Geneva

UNCTAD (2007) World investment report 2007. In: United Nations conference on trade and development, Geneva

UNCTAD (2011) Global and regional FDI trends in 2010. Highlights. In: United Nations conference on trade and development, issue no. 5, Geneva

UNCTAD (2012) World investment report 2012. In: United Nations conference on trade and development, Geneva

UNEP (2007) Global environment outlook GEO4—Environment for development. United Nations Environment Programme, Nairobi

WWF (2001) Preliminary assessment of the environmental and social effects of liberalization of tourism services. WWF international discussion paper, Geneva, May

Chapter 2
The Main Features of the FDI Phenomenon

2.1 FDI Definition and Players

According to the definition given by the UNCTAD, FDI is an investment involving a long-term relationship and reflecting a lasting interest and control by a resident entity in a given economy (foreign direct investor or parent enterprise) in an enterprise resident in an economy other than that of the foreign direct investor (FDI enterprise or affiliate enterprise or foreign affiliate) (UNCTAD 2007: 245).[1] As can be observed, two keywords represent the main feature characterizing the definition: lasting interest and control. In fact, a FDI is normally distinguished by the other form of private capital, and particularly from the portfolio equity investment, because it implies long term investment relationship while the latter results more volatile. With regard to the second feature of control, the identification of a FDI by general convention occurs when a minimum of 10 % shareholding in a foreign firm's capital is considered.

FDI is an activity which is normally run by Multinational Corporations (MNCs). In fact, the literature generally refers to MNCs as those firms which undertake FDI as the main motivation of their activity. No single definition of what a MNC is exists.

---

[1] Other definitions are provided by the International Monetary Fund (IMF) and the Organization for the Economic Cooperation and Development (OECD). The IMF defines FDI as an investment made to acquire a *lasting interest* in an enterprise operating in an economy other than that of the investor, the investor's purpose being that of having an effective voice in the management of the enterprise (IMF 1993: 93). The OECD defines FDI as the category of international investment that reflects the objective of a resident entity in one economy to obtain a lasting interest in an enterprise resident in another economy. Although all these definitions are slightly different from that given by the UNCTAD, they do not show significant changes especially with regard to the basic features of FDI.

P. Pazienza, *The Relationship Between FDI and the Natural Environment*, SpringerBriefs in Economics, DOI: 10.1007/978-3-319-04301-2_2,

However, a basic distinction is usually made between International Firm (IF), Multinational Corporation (MNC) and Transnational Corporation (TNC).[2]

The actual implementation of FDI may take either the form of greenfield investment, or the form of cross-borders Merger and Acquisitions (M&As), or the form of Joint Ventures (JV). By definition, the first form refers to an investment made "from scratch", aimed at creating a completely new enterprise in host territorial areas where no previous production, distribution or other facilities exist. This type of investment can be very costly for the investor, but it is well received by host countries, because of its high job-creation potential and its relevant capability to increase the value-added of the host country's production. As the name implies, M&As are typically implemented via the ownership change of existing enterprises. It specifically refers to investment dealing with the buying, selling and combining of companies. This mode of investment has the advantage of being cheaper than greenfield investment and gives the investor quick access to the market of the host country. Lastly, JV investment is made by a foreign firm under an agreement with one or more firms or government institutions in the host country, as well as other companies outside the host country. All parts in the agreement are committed to bringing their own skills and expertise to the investment operation such as, for example, the knowledge of the local or national market and bureaucracy, technical expertise, financial capability, etcetera (Moosa 2002).

2.2 Classification of FDI

The classification of FDI typically distinguishes the operational view of the source country of the investment from that of the host country. From the view of the source country, or the investor's view, FDI can be categorized in horizontal, vertical and conglomerate. Horizontal FDI refers to an investment operation aimed at the horizontal expansion of the production. This means that an investor, operating in the source country, decides to produce abroad—in a country which will host the investment—the same or a similar type of product he produces at home with the aim of expanding his market opportunity. What characterizes horizontal FDI is the lack of product differentiation between that produced at home and that in the host country. This kind of investment is typically run to exploit the advantage of a certain power position in the market (i.e. monopoly or oligopoly) a

[2] An international firm can be defined as a firm which works in importing and exporting goods produced in the domestic market and then exported abroad and vice versa. The evolution of an international firm can bring to the identification of a multinational firm, which refers to a firm producing both at home and abroad (through subsidiaries, affiliates and joint ventures). The further evolution of he firm can identify the transnational corporation. This occurs when a firm evolves at such a point that difficulties arise as to the identification of its home country (Moosa 2002).

firm derives from holding, for example, patents and where the expansion in the home country may contravene anti-trust regulations. Vertical FDI is, instead, undertaken with the aim of gaining the economic advantages that an investor derives from a better management of his organizational chain. In fact, he may consider it advantageous to be as close as possible to the market of raw materials acquisition and/or to final consumers. The earlier case may occur through investment to buy other firms working as raw materials suppliers (backward vertical FDI). The latter may take place through the acquirement of distribution outlets (forward vertical FDI). Lastly and very simply, conglomerate FDI represents a mix of the previous two types (Moosa 2002).

From the view of the host country, FDI can be categorized into import-substituting, export-increasing and government-initiated FDI. Import-substituting FDI is basically determined by aspects such as the market size of the host country and the existence of transportation costs and/or trade barriers. It refers to an investment which enables the host country to become producer of certain products which were previously imported. As a consequential result, imports by the host country, but also exports by the source country, will decline with a potentially realistic improvement of the balance of payments of the earlier. Export-increasing FDI occurs when a country becomes object of interest of an investor, who seeks further or new sources of input factors. In such a case, the host country increases its export of certain products (normally raw material and/or intermediate goods) to the investor's country and/or other countries where his subsidiaries are located. Government-initiated FDI refers to that form of investment which is stimulated by the provision of forms of incentives offered by governments to attract investment in the attempt to improve their balance of payments conditions.

A possible last classification of FDI distinguishes between expansionary and defensive FDI. Expansionary FDI is a form of investment which is aimed at the exploitation of firm-specific advantages (e.g. scale effects, R&D intensity, profitability and technology acquisition, etc.) in the host country and has the additional benefit of contributing to the growth of sales of the investing firm both at home and abroad. Defensive FDI is that investment which is aimed at reducing production costs and, in doing so, seeks cheap labour (or other cheap input factors) in the host economy (Chen and Ku 2000; Chen and Yang 1999).

2.3 The Measures of FDI

With regard to the quantitative aspect of FDI, it can be observed how its measure is generally expressed either in terms of flow or in terms of stock. FDI flows include the capital invested—either directly or indirectly through related enterprises—by a foreign investor in an enterprise, or the capital received from an enterprise by a foreign investor. More specifically, "for associates and subsidiaries, FDI flows consist of the net sales of shares and loans (including non-cash acquisitions made against equipment, manufacturing rights, etc.) to the parent company plus the parent firm's share of the affiliate's reinvested earnings plus total net intra-company loans (short- and long-term) provided by the parent company. For branches, FDI flows consist of the increase in reinvested earnings plus the net increase in funds received from the foreign direct investor. FDI flows with a negative sign (reverse flows) indicate that at least one of the components in the above definition is negative and not offset by positive amounts of the remaining components". With regard to FDI stocks, we can learn how "for associate and subsidiary enterprises it represents the value of the share of their capital and reserves (including retained profits) attributable to the parent enterprise (this is equal to total assets minus total liabilities), plus the net indebtedness of the associate or subsidiary to the parent firm" (UNCTAD).[3] Furthermore, it is important to highlight how FDI flow and stock may take the form of inward or outward investment depending on the direction it takes. Very simply, FDI flow or stock is inward when an investor of a foreign country invests in a considered country. It is outward when an investor from a home country invests abroad (Moosa 2002).

According to Cantwell and Bellak (1998), it is generally referred that the practice of reporting FDI in terms of stock is basically unsatisfactory. Stocks are expressed in terms of their "book value", namely their historical cost, which does not take into consideration their age distribution and makes international comparison almost impossible. Apart from this specific aspect, we must understand that measuring FDI is not straightforward because of the existence of problems especially occurring when the investment takes the form of machinery or contributions of technological capitalization. Furthermore, due to the reluctance of most countries to provide comprehensive information on the foreign operations of their companies for reasons of secrecy, gaps exist in FDI statistics available for source and host countries (Moosa 2002).

2.4 The Effects of FDI

After having insofar referred the main featuring aspects of FDI, we can now move onto giving a broad look at the effects it generates. The FDI dynamic involves the transfer of various elements (financial capital, technology, labour skills, etc.) from a country (the source of the investment) to another (the destination or recipient of the investment). This process implies the rise of costs and benefits for the countries involved. Due to the existence of a general disagreement—based on the existence of different views pro and con the globalization phenomenon—it is not really clear what costs are endured and what benefits are gained by the countries. This is

[3] These definitions are gathered from the UNCTAD web-page in the source and definition section available at http://unctad.org/en/Pages/DIAE/FDI%20Statistics/Sources-and-definitions.aspx.

particularly true from a quantitative view. However, the FDI effects issue is basically treated from the host country's point of view. According to Moosa (2002), the effects of FDI on an investment host country can be of the following type: economic, political and social. In short, the social issue mainly concerns the creation of enclaves and foreign elite in the host country, as well as cultural and behavioural changes as a consequence of a sort of "contamination" resulting from the contact between the foreign and local entities. The political effects refer to the question of national sovereignty. It is natural to think that—and this could be particularly true in Less Developed Countries (LDCs)—because of the relevance of the interests implicated by the management of a MNC, a threat for the national political autonomy of the host country could exist. The economic effects are distinguished in macro and micro effects as shown in the scheme below.

The earlier are often referred to in the same terms as a rise in foreign borrowing. If there is unemployment and capital shortage—that is the typical case of LDCs—FDI (which is provision of capital) leads to an increase of output and income together with a reduction of unemployment in the host country. In this sense FDI has a beneficial effect on the balance of payment, but its effect in terms of trade is indeterminate since this will depend on whether the impact of increased output falls on import substitutes of export. The micro effects, instead, concerns structural changes in the economic and industrial organization. Broadly speaking, they refer to individual firms and individual industries, particularly those exposed and associated with FDI. Within this context, for example, a relevant argument is whether FDI leads to a more competitive economic environment. It must be highlighted, however, that the issue of the economic effects of FDI has very often failed to consider those associated to the natural environment. At least this is true up to the late 1990s when the sensitivity for this specific aspect began to appear in the reports of some international organizations. We have inserted the environmental component in the Fig. 2.1 as an adaptation of the discussion of the already-mentioned author, who has the merit of having summarized the state of art of the literature previously produced.

With specific regard to the relationship between FDI and the environment, we must highlight how it can be considered as a subset of the FDI literature and, especially in the last decade or so, it has been abundantly treated by academicians, politicians and experts. This has generated a massive production of writing, whose analysis is not straightforward. From a methodological point of view, two basic ways of empirically analysing the issue can be observed in the literature. The first looks at the statistics of FDI flows, environmental data and information on the environmental regulatory systems of various considered countries or areas in the attempt to identify the existence of some linkages between the two aspects. The second considers the behavioural aspect of firms to understand how they make their investment location decision and if environmental factors play a role in this process. Relevant information on these issues can be found in various UNCTAD reports and especially in two of them (UNCTAD 1999, 1993). The first, which is still now generally recognized as the most comprehensive study of the environmental performance of MNCs, basically refers that larger companies are more

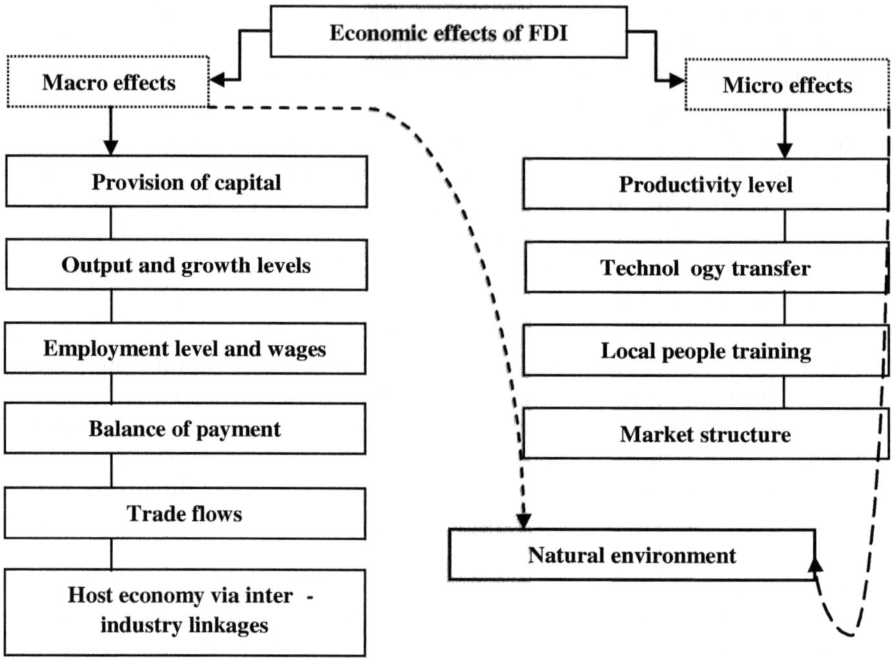

Fig. 2.1 Economic effects of FDI. *Source* built and adapted from Moosa's (2002) discussion

likely to have better management performance. The second contains a useful update of the previous and a valuable discussion on the environmental effects of FDI in emerging economies. Additional information can be found in some other useful works produced by the OECD, where the research developed is categorized into four macro themes, which we will refer to with the aim of presenting a methodologically clear discussion of the specific literature. However, for the convenience of our discussion, we group the issues analysed on the FDI-environment relationship into three thematic areas: (1) the environmental effects of FDI flows; (2) the competition for FDI and its effects on environmental standards; (3) the cross-border environmental performance (OECD 2002a, b, 1997).[4] The

[4] A fourth area can also be considered, which is represented by the regulatory impact of investment rules. Although this is referred to be the most recent area of work, so far it has basically focused on understanding whether or not investment and environmental protection are pursuable as a common achievement. Furthermore, most of the work carried out in this field of discussion has taken into consideration the analysis of investment models and agreement regulations such as those in the North American Free Trade Agreement (NAFTA), the OECD Multilateral Agreement on Investment (MAI), and Bilateral Investment Agreements (BITs) (OECD 2002b). Since the rationale of the studies conducted in this field seems more focused on the analysis of the juridical content of agreements (e.g. Ignacio 2003) and takes a different direction from the scope of our work, we purposely fail to report a more extensive description of it in the conviction that a sufficient note can remain in these few lines.

next chapters will focus on this framework to build up a better understanding of the relationship between FDI and the natural environment.

2.5 Some Concluding Considerations

In this chapter we have highlighted the main aspects and facts of the FDI phenomenon. By referring to the technical literature, we have analysed its definition and classification, and understood its measures and players. In addition, we have briefly observed the economic, political and social effects it generates in investment host or receiving countries. For the specific purpose of this work, however, a more detailed attention has been paid to the first type of effects. The analysis of those works focusing on the economic effects of FDI highlights how the natural environment (which can be broadly seen as a supplier of input factors of production and a receiver of waste and residuals from production processes) is very often left unconsidered. As has been commented, this is particularly true up to the late 1990s when a renewed sensitivity for an environmental reflection on issues such as trade and investment began to appear in the reports of major international organizations. The analysis of the literature on the FDI-environment relationship produced insofar is not straightforward because the quantitatively relevant production of the last decade is very fragmented. As has been said in the previous sections, however, the works produced can be methodologically grouped into three main veins of discussion: (1) the environmental effects of FDI flows; (2) the competition for FDI and its effects on environmental standards; (3) the cross-border environmental performance. In consideration of these three discussion areas, the following chapter will present an updated literature review of these issues which are the core subject of our argument.

References

Cantwell J, Bellack C (1998) How important is foreign direct investment? Oxford Bull Econ Stat 60:99–106

Chen JR, Yang CH (1999) Determinants of foreign direct investment: comparison between expansionary FDI and defensive FDI. Taiwan Econ Rev 27:215–240

Chen TJ, Ku YH (2000) The effect of foreign direct investment on firm growth: the case of Taiwan's manufacturers. Jpn World Econ 12:153–172

Ignacio M (2003) Foreign direct investment and the protection of the environment: the border between national environmental regulation and expropriation. Eur Environ Law Rev 12(3):70–82

IMF (1993) Balance of payments manual. International Monetary Fund, Washington, D.C

Moosa IA (2002) Foreign direct investment: theory, evidence and practice. Palgrave MacMillan Ltd., New York

OECD (1997) Foreign direct investment and the environment: an overview of the literature. In: Document of the negotiating group on the multilateral agreement on investment (MAI). OECD, Paris, France

OECD (2002a) Environmental issues in policy-based competition for investment: a literature review. In: Document of the working party on global and structural policy. OECD, Paris, France
OECD (2002b) Environmental benefits of foreign direct investment: a literature review. In: Document of the working party on global and structural policy. OECD, Paris, France
UNCTAD (1993) World Investment Report 1993. United Nations Conference on Trade and Development, Geneva
UNCTAD (1999) World Investment Report 1999. United Nations Conference on Trade and Development, Geneva
UNCTAD (2007) World investment report 2007. United Nations Conference on Trade and Development, Geneva

Visited Web-Pages

http://unctad.org/en/Pages/DIAE/FDI%20Statistics/SOurces-and-definitions.aspx

Chapter 3
The Environmental Effects of FDI

3.1 The Classical Analytical Approach

The first vein of discussion, related to the environmental effects of FDI, is claimed to be one of the research areas where the literature lacks a better and more appropriate scientific understanding (OECD 2002b). Even at the time of writing, this research field still appears largely unexplored since—as has been already mentioned above—a great deal of the scientific work developed insofar has focused more on the environmental features eventually influencing the location of the investment decision of firms and the countries' environmental regulatory competition for FDI rather than on this aspect. Although some works have been done to cover the gap in this thematic area, a strong call for further research still exists. This body of discussion comments on how FDI can generate benefits and costs or, which is likely the same, opportunities and risks. This is particularly true for host or receiving countries of FDI and their communities. For example, FDI may boost economic growth, generate structural efficiency together with other positive effects, but it can also generate environmental degradation. Indeed, FDI can spread industrial activity, stimulate the production and consumption of industrial polluting goods, all this resulting in an increase of the sources and forms of pollution. However, it is also often argued—and this has been proven to be more than a simple hypothesis—that foreign investors bring new technologies to receiving countries. This would enable receiving countries to implement environmental protection projects and actions. In this perspective, the economic expansion driven by FDI may also generate a generalized improvement in the environmental sphere. For this reason, some authors point out that FDI should be considered neither a boon nor a bane. It is both, since the wide differences of locations, sectors and investors involved in FDI allow us to find evidence to support both views (i.e. Gentry 1999). Indeed, depending on each specific circumstance, examples of the existence of environmental positive or negative marginal effects of FDI can be found in some countries and not in others. With regard to this, the literature claims a lack of studies able to identify the "net effect"

P. Pazienza, *The Relationship Between FDI and the Natural Environment*, SpringerBriefs in Economics, DOI: 10.1007/978-3-319-04301-2_3,

of FDI on the environment.[1] However, this might be seen as a complex—or even impossible—job to do, because of the existence of two constraints. The first refers to the extreme difficulty in discriminating the effects of the activity run by domestic industries from those of foreign firms or their affiliates. Furthermore, FDI does not occur as an isolated phenomenon, which affects the environmental sphere only, but it also interacts very strictly with other linked factors. For this reason most of the studies on the environmental effects of FDI are carried out by decomposing them into scale, composition (or structural) and technique (which is also associated to a technological aspect) effects. Very briefly, while scale effects refer to the environmental results of the expansion of the economic output, composition and technique effects respectively refer to the change of the industrial structure of an economy (due to a reallocation or reorganization of the production and consumption structure) and to changes of the production methods associated to the development and diffusion of technology (OECD 2002b).

3.1.1 The Scale Effects

More specifically, scale effects would refer to the impacts that the increment of an economic activity—arising from the entry of new foreign investment to be consistent with the context of our discussion—generates on some considered environmental features of host countries. With regard to this, the issue related to the practicability of an economic growth process "decoupled" from environmental degradation comes out in all its evidence. As already referred in a footnote in the introductory section, the recognition that "decoupling" is not a realistic way of approaching an environmentally-sound growth induces us to think that economic growth always generates a more or less relevant, however, negative impact on the environment (Jackson 2009). In fact, these types of effects are normally and generally expected to be detrimental and negative to the environment and can be easily understood if the case of resource-intensive industry sector is referred to. Particularly in this sector, the increase of production requires the extraction and the

[1] Although this approach of analysis is often said to be desirable, the difficulties implicitly existing in its performance would impede the achievement of meaningful results. As reported by UNCTAD, there are plenty of studies which have come to the same conclusion in demonstrating that specific industrial activities (such as those related to production in the sectors of chemical and allied, pulp and paper, mining for mineral and iron, cement, glass and ceramics) may result highly pollutant. However, because of data insufficiency, research has failed to prove the existence of precise relations between FDI flows and the potential pollution intensity of these considered industry sectors. Hence, the report highlights that the "net effects" of FDI on the environment depend on a combination of macro and micro aspects. The first can be related to the FDI profile such as, for example, the type of industry sector in which it takes place, and the extent to which it involves pollution-intensive activities. The micro aspects could refer to specific decisions with regard to the management of their production activities and the adoption and diffusion of environmentally sound technologies (UNCTAD 1999: 294–295).

use of more resources and the generation of more waste (O'Connor 2000). The existence of a debate can be further observed with regard to the determination of the size of the scale effects (Nordström and Vaughan 1999), which is followed by some clarifying works stating that this basically depends on the environmental feature one investigates (e.g. OECD 2001). However, a different view is expressed in other works where it is referred that the detrimental result of the scale effects on the environment can be alleviated, or even counterbalanced, in a longer time span because of the existence of an "inverted-U" relationship—also known as Environmental Kuznets' Curve (EKC)—between some features of environmental quality and economic growth. Evidence in this sense has been produced by various relevant works in the past couple of decades.[2] Hence, according to this view, countries with an economic growth induced by FDI inflows may experience a deterioration of their environmental quality—due to the increase of their pollution caused by the boost of industrial activities—at the beginning of their process and up to a certain stage. Afterwards, pollution levels should start to decline because of the increase of a demand for environmental quality taking place in association with the increase of the economic welfare of the population.[3] However, evidence proving the inexistence of the EKC are also extensively reported in the literature (e.g. Stern 2004a, b; Perman and Stern 2003; Yandle et al. 2002). Having reported all this, however, it must be stressed that for the reasons already mentioned above, a wider debate on the FDI-environment relationship cannot be identified. The issue suffers from the lack of data and analysis and is almost basically limited to the concepts and reflections referred. The literature analysis also highlights that, due to the difficulty in conducting studies for the straight identification of net effects

---

[2] For example, some works found a positive relationship between environmental regulation and the growth in the population per-capita income (Dasgupta et al. 2001). Other works verified how many indicators of environmental quality deteriorate at an initial stage of economic growth, but they improve when economic welfare overtakes certain thresholds (Grossman and Krueger 1993a, b, 1995; Selden and Song 1994; Shafik 1994; Shafik and Bandyopadhyay 1992). More recently, an analysis of air and water pollution for a set of 120 countries between 1960 and 2001 observed the existence of the EKC for water only (Gassebner et al. 2011).

[3] The achievement of a detailed survey of the EKC literature is certainly not the scope of our discussion. However, it seems the case to highlight that a series of limitations arise to cool the optimistic feeling which may be generated by approaching the view expressed by the EKC issue. As reported by authors who have profoundly characterized this scientific debate, in many countries the "turning point" of the inverted-U curve could be found at quite a high level of their population's per-capita income, this opening the way to the fear of irreversibility of environmental degradation meanwhile generated (Panayotou 2000, 1997; Opschoor 1995). With regard to this, it has more specifically been observed how an EKC might exist for both local and global pollutants (e.g. Grossman and Krueger 1995; Seldon and Song 1994). In this respect, it must be highlighted, however, that the EKC shows its validity only for some pollutants, for some countries and not all the time (Stern 2004a, b; Munasinghe 1999; Barbier 1997). Generally speaking, it has also been observed that local pollutants respond better than global ones to investigations of the EKC hypothesis in the sense that the turning point of the earlier are lower than the latter (Meers 2000). Finally, some empirical evidence of the linkage between economic growth induced by trade and environmental deterioration exists, but this does not mean the same relationship can be validated for FDI-induced growth (Dessus and Bussolo 1996).

between FDI and the environment, researchers changed their analytical approach at a certain point and came to focus more on case studies—many of which are from Asian countries—with the aim of understanding the environmental impact and the management strategies of foreign firms. Even these analyses do not definitively resolve the debate, since evidence of positive as well as negative environmental effects is found (Guoming et al. 1999; Jha 1999). More recent evidence can be tracked in other works which investigate the metal mining sector of some countries of the Sub-Saharan African region (Ghana, Tanzania and Zambia), where MNCs are found to cause both negative and positive effects on the environment. While the negative effects can be mainly referred to as deforestation, air, water and dust pollution, the beneficial aspects are represented by the adoption of better management practices and the introduction of environmentally-friendly technologies, these giving rise to positive technological effects in the sense that will be disclosed later in the paragraph (UNCTAD 2007; Boocock 2002).[4]

3.1.2 The Composition Effects

Broadly speaking composition (or structural) effects are associated with the adjustment within and between economies when a shift occurs in the pattern of economic activity, including a shift from one production sector to another or from one product to another, changes in the price of input factors and final products, changes in industry ownership, changes in efficiency. A more focused view on the environmental consideration would see a structural effect when a shift in the pattern of resource use occurs (OECD 2001). On the assumption that trade and investment liberalization encourage allocative efficiency among countries, these effects are expected to have a positive impact on the environment.[5] According to the last available UNCTAD report, FDI flows have experienced structural shifts which can be observed at geographical and sectorial levels.[6] With particular regard

[4] More specifically, some of the mentioned studies also refer that in the attempt to offset such degrading situations, MNCs have introduced more environment-friendly technologies and higher standards of environmental protection in comparison to the local firms working in the same sector.

[5] It is the case to highlight that, in agreement with some authors, the occurrence of structural changes clearly leads to some shifts in the optimal allocation of resources. However, it is not clear if this implies the fulfillment of ecological sustainability constraints. Hence, the identification of a strict and virtuous relationship between economic efficiency and ecological sustainability cannot be a natural expectation (Common and Perrings 1992).

[6] From a geographical point of view, in the past decade the pattern of the FDI dynamic has changed enormously. New world areas have come to the spotlight as host and receiving countries. Shifts in the patterns of bilateral FDI relationships can also now be observed among developed countries, and between developed and emerging economies. The analysis of recent trends shows a significant increase of FDI flows from developing and transition economies and in the South-South relationships (UNCTAD 2007: 19). From a sectorial point of view, it can be appreciated

to the sectorial shift, the arising question highlighted in the literature under analysis tries to understand to what extent the FDI shift towards the service sector can result beneficial to the environment. Some approaches in the literature support this hypothesis. For example, they argue that newly industrialized countries can move from the primary to the service sector passing through a low polluting light-manufacturing production experience. This should allow them to actually jump phases of heavy industrialization, such as those experienced by the traditional industrialized countries during certain phases of their economic history, thus gaining significant environmental advantages (OECD 2001; Gentry 1998).[7] Another analysis states that the service sector is the one where the implementation of better environmental practices is more likely to happen. It supports this hypothesis by referring to the hotel sector as an example in which FDI induced practices for a more efficient use of water, energy and waste are more widely put into action, thus generating positive environmental effect (UNCTAD 1999). As a counter fact, a case study by WWF (2001), while recognising the beneficial effect of FDI in tourism activities, also stresses its possible negative effect on the environment. The benefits can be particularly seen in terms of income generation, which also allows the creation of infrastructure and facilities for environmental protection. The negative aspects may be related to the ownership structure of tourism service facilities. In fact, when the tourism sector is driven by FDI—and this is the case in many developing economies—this would imply that the economic benefit will almost always flow out of the FDI host country, where it was generated, while environmental costs generated by the running of the activity and tourist use of the territory will remain upon the shoulders of the local communities. Another question is also posed by the literature and refers to the understanding of whether or not the world economic structural shift in favour of the service sector will push manufacturing industries to move from developing to rapidly industrializing countries, thus giving rise to some negative environmental implications (O'Connor 2000). Although research is claimed on this aspect, available empirical

(Footnote 6 continued)
how over the past 25 years FDI has grown notably in absolute terms in the three main economic sectors (primary, manufacturing and services). However, the observation in terms of stock shows the primary (particularly referred to the natural resource sub-sector) and manufacturing quotas have considerably decreased. Meanwhile, a significant shift in favour of the service sector can be appreciated. In terms of FDI stock, in fact, while the primary sector in 2005 represents one tenth of the total, that is a slight decrease with respect to the figure in 1990, the manufacturing sector accounts for 30 % of the total in 2005 against 41 % in 1990. The FDI stock in the service sector performs a significant increase arriving to represent 61 % of the total in 2005 against 49 % in 1990 (UNCTAD 2007: 22).

[7] With regard to this, however, a reflection should be made with regard to the different nature of the many activities characterizing the service sector (e.g. the difference between financial and air transport services) and to the different environmental impacts they can generate. This considered, further research is claimed for a better understanding of the wide variety of environmental impacts which can arise from the service sector activities (OECD 2002b).

evidence shows that FDI flows to the service sector of developing countries is an increasing phenomenon and, on the basis of the consideration of some data, it seems supported not only by service-based MNCs but also by the establishment of others operating in the other two sectors (UNCTAD 1999, 2007).

3.1.3 The Technique Effects

The aspect of technique (or technology) effects refer to the development, transfer and diffusion of technology as a phenomenon associated to the movement of international investment flows. This can happen through the transfer of physical goods (i.e. capital goods) and the transfer of tacit knowledge (UNCTAD 1999). In a free trade and investment context, technology effects would normally be expected to exert a positive (or at least neutral) impact on the environment, since the efficiency concept associated to them implies that goods are produced with lower labour and capital inputs, which as a result also means a decreasing impact on the natural resources system. In other terms, technology innovation is generally felt to play an offsetting role by bringing more resource-use efficiency (OECD 2001). However, this cannot be considered the rule of thumb. Other views referring to an opposite perception exist. Hence, two different ways of thinking—that is a double hypothesis—characterize the technology effects induced by investment activity. In some situations, foreign investments can certainly generate positive spillover in the environmental sphere of the investment host country, as a result of the use of environmentally-friendly technologies in producing goods and exploiting resources. In this direction goes, for example, the evidence produced in some works already mentioned. In analysing the technological side of the Chinese FDI inflow, the study refers that foreign investors have introduced better technologies and more appropriate environmental structures and practices which were inexistent in the country before (Guoming et al. 1999). Other works suggest that the presence of MNCs working with new technologies in host countries can also represent a push or an incentive for national and local firms to implement similar production methodologies. As has been observed, in a comparison with more advanced and better organized MNCs, domestic firms may feel an incentive to imitate their production schemes. In this way a so-called "reverse engineering" process starts, although it can be strongly conditioned by the more or less rigidity of the property rights system (Panayotou 2000; Blömstrom and Kokko 1996; Coe and Helpman 1995).[8] Similar empirical evidence is also produced by a relatively

[8] Furthermore, the study by Blömstrom and Kokko (1996) highlights that the presence of multinational firms in a country seems to generate further technological spillover among the supplier industries, which can be beneficial to the environment. In fact, by requesting factor inputs (e.g. raw material, component parts, etc.) with specific quality standards and furnishing supplier firms with appropriate technical aid to achieve the required standards, multinationals can encourage these firms to improve their technological performance. For other aspects, the same

recent work which analyses the Vietnamese investment context in the manufacturing sector while exploring—among other aspects irrelevant to the purpose of our discussion—whether or not FDI generates vertical or horizontal technology spillover on domestic firms through the use of a panel data technique. The study concludes by confirming that FDI is found to be a relevant tool to improve production efficiency and expand the small and inexperienced domestic enterprise of the considered sector (Nguyen et al. 2008). Conversely, in other situations technology may have a negative impact on the environment by playing a detrimental effect. This can happen, for example, when the object of the technological transfer is either archaic and obsolete equipment or machinery or technology banned in the FDI source country due to the previous knowledge of its negative environmental effects (OECD 1997). In a small number of cases, evidence of this "technology dumping" is reported with regard to some cases of companies which were dismantling obsolete production plants in industrialized countries to move them to emerging economies (Esty and Gentry 1997). Similar evidence is reported with regard to the Chinese leather, footwear and plastic sectors, where in the past Asian investors were seen to implement very poor environmental performance, thus negatively affecting the environment and the health of the local communities (Guoming et al. 1999). Yet another example comes from Malaysia, where observations were made of the very low environmental practices implemented by some MNCs in the chemical and copper mining sectors and in the disposal of radioactive residuals of production processes (Rasiah 1999). At other times, it is found that technology implemented in foreign investors' firms does not play any relevant role in generating beneficial spillover to domestic firms of host countries. An investigation related to the Estonian transitional economy on the existence of spillover from technology transfer to the domestic firms and the relationship between this spillover and the capacity of domestic firms to absorb them shows how spillover from technology transfer depends on a number of aspects, such as the size of the receiving firm, its trade orientation and its ownership structure. The study finds that small, non-exporting and foreign firms gain a higher beneficial effect from spillover than domestic firms are able to do. In fact, in contrast to the expectations, domestic firms do not enhance their ability to attract the beneficial aspects of technology spillover by failing to catch up with foreign firms in most industries, thus failing to bring in a higher level of efficiency in their production schemes (Sinani and Meyer 2004).

The "rebound" or "take-back" effect[9] is a further argument supporting the view that technique effects do not play a beneficial role on the environment. Its current debate predominantly focuses on what is generally known as the "Jevons'

(Footnote 8 continued)
study also reports the observation of other beneficial spillovers, which can arise when people and experts previously employed in international firms—especially when they are the subject of significant training programmes—are successively engaged by national and local firms.

[9] This type of effect is normally expressed in terms of a ratio between the lost environmental benefits and the expected environmental benefit (Murray 2013).

paradox", that is the possibility of a back-fire occurring as a result of technology improvements in energy efficiency. More specifically, the "rebound" effect refers to the fact that technology improvements are mainly implemented to pursue reductions of the levels of energy use and carbon emissions. However, it has been considered and also observed that efficiency gained in energy consumption thanks to technology advances normally results in the reduction of the per-unit energy prices. This would unavoidably determine an increase of energy consumption and that of polluting emissions with the consequent result of partially offsetting the efficiency gain in energy use (Greening et al. 2000). As has recently been observed, for some authors this issue appears to be just a theoretical exercise. Other authors, instead, empirically investigate the issue in question while focusing on the specific features of a given technology. Apart from this, however, the relevance of the "rebound effect" issue is associated to the policy question regarding the identification of the scale at which consumption pattern changes might occur, at a considered technology level, to ensure a beneficial outcome for both the environment and the economy (Murray 2013; Burns and Potts 2011).

3.2 More Recent Analysis Approaches

Moving away from those works specifically related to the analysis of each single effect we have just mentioned, the most recent literature shows how researchers have come to follow new approaches to investigate the FDI-environment relationship. This vein of research is richer in works focusing on the trade-environment relationship rather than the FDI-environment one.[10] Apart from this, however, the new approaches consist in developing analysis while considering the empirical evidence as associated to all these aspects (scale, structural or composition and technique or technology effects) together. For example, while focusing on the trade issue, some authors consider scale, technique and trade-induced composition effects while using Sulphur Dioxide (SO_2) data for 43 countries over the period between 1971 and 1996. Their empirical result shows consistently higher elasticity of technique effect over scale effects. In their study, trade-induced composition shows a generation of positive consequences for the environment. They conclude by saying that free trade is good for the environment (Antweiler et al. 2001). This result is partially supported by a complementary study by Cole and Elliott (2003), who assess a combined scale and technique effect for SO_2, Nitrogen Oxides (NOx), Carbon Dioxide (CO_2), and Biological Oxygen Demand

[10] We also make reference to the trade issue, since we are aware of the fact that FDI and trade can be intended as the two faces of the same coin. It is generally recognized that the perception, which has been empirically proven in most cases, that trade and FDI are interlinked in various modes and they are two ways—sometimes alternatives, but increasingly complementary—of servicing foreign markets (ie. Chaisrisawatsuk and Chaisrisawatsuk 2007; Hejazi and Safarian 2001; Baldwin 1994).

(BOD). They find technique effects are dominating scale effects for SO_2 and BOD with scale effects dominant for NOx and CO_2.

With more specific regard to the FDI-environment relationship, for example, a work aimed at understanding whether or not FDI is harmful to the Chinese environment finds a negative relation between the FDI inflow and the air quality measured in terms of SO_2 emissions, thus stating that FDI generates a beneficial effect to the environment of the host country. Benefits happen thanks to a process of technological innovation (the technology effect), implicitly associated to the foreign investment dynamic, which brings higher levels of production efficiency and pollutant abatement as a result (Liang 2006). A different view is referred by the evidence produced by another author and his panel data model built for the period between 1994 and 2001 with regard to 29 Chinese provinces affected by industrial SO_2 emissions. He observes a positive relation between the inward FDI stock and the emission levels of SO_2: a 1 % increase in the FDI stock generates a 0.098 % increase in SO_2 emission levels. He relates this overall very small negative impact of FDI on the environment to the fact that the country produces a relatively higher pollution efficiency level as a result of the technology effect and to a composition effect, which is heavily influenced by the inflow of foreign capital in searching for lower compliance costs of pollution regulation (He 2006). Another similar view is expressed in a more recent work, where the FDI-environment relationship is investigated over the period between 1985 and 2006 for 110 developed and developing economies. Through the use of the econometric technique of panel data, the study finds a significant and positive linear relationship between the two considered aspects of the flow of foreign direct investment and energy emissions considered in terms of CO_2. The authors conclude by saying that the increase of FDI generates increases in the levels of environmental degradation (Shahbaz et al. 2011). Through the use of a panel data set of 29 Chinese provinces for the period between 1992 and 2004, another study investigates the effect of FDI on the emission levels of five different pollutants and assess the technique, scale and composition effects. The analysis result shows that, although FDI contributes to the reduction of pollution emissions in the whole of China, the FDI environmental impact varies significantly among different regions and pollutants (Bao et al. 2011).

As can be observed, these works—as the majority of the research carried out in this thematic context—focus their attention on the FDI-environment relationship while working on aggregated data and disregarding the specification of the activity sectors, which in our view should be taken into consideration for a more thorough investigation. In this direction another recent work investigates French data of the FDI outflow, disaggregated at sectoral level, which reached a mix of developed, emerging and developing countries between 1999 and 2003. Through the use of a simultaneous equations model and among other analysis targets, it also assesses the FDI impact on the environment (this considered in terms of CO_2 and BOD emissions) of host countries. The result shows a positive relationship between the FDI outflowing to the manufacturing sector of host countries and CO_2, this showing the existence of a carbon leakage dynamic. A different result is achieved, instead, with regard to the relationship between FDI outflowing to the manufacturing sector of host countries and BOD water emissions which shows an inverse relationship (Ben Kheder 2010).

3.2.1 Some Concluding Considerations

The examination of this part of the literature on the environmental effects of FDI does not give us a clear understanding of whether or not FDI affects the environment positively, negatively or neutrally. Results are specifically related to the context of the analysis where they are achieved. As observed, in fact, a wide part of the scientific debate and the production of analyses as a result concentrate on only one dimension of the three different categories (either scale, or structural composition, or technology effects) of the FDI-environment relationship. Apart from some more recent works, studies still seem to fail to consider these three aspects all together. For this reason, a call for further research in this area is still generally made. Furthermore, according to some ideas, an interesting point which does not seem to be properly addressed yet is the understanding of whether the structural shift most countries are experiencing by moving from the manufacture to the service sectors is environmentally valuable (OECD 2002b).

References

Antweiler W, Copeland BR, Taylor MS (2001) Is free trade good for the environment? Am Econ Rev 91:877–908

Baldwin RE (1994) The effects of trade and foreign direct investment on employment and relative wages. OECD Econ Stud 23:7–54

Bao Q, Chen Y, Song L (2011) Foreign direct investment and environmental pollution in China: a simultaneous equation estimation. Environ Dev Econ 16(1):71–92

Barbier E (1997) Introduction to environmental Kuznets curve special issue. Environ Dev Econ 2(4):369–381

Ben Kheder S (2010) French FDI and pollution emissions: an empirical investigation. CERDI working papers. Clermont-Ferrand, France. Available at http://cerdi.org/uploads/sfCmsContent/html/323/BenKheder.pdf

Blomstrom M, Kokko A (1996) Multinational corporations and spillovers. CEPR discussion paper no. 1365, London, UK

Boocock CN (2002) Environmental impacts of foreign direct investment in the mining sector in Sub-Saharan Africa. In: Foreign direct investment and the environment: lessons from the mining sector, OECD, Paris, France

Burns C, Potts M (2011) The "rebound effect": a perennial controversy rises again. eSolutions J 4(2). Available at http://www.rmi.org/

Chaisrisawatsuck S, Chaisrisawatsuck W (2007) Imports, exports and foreign direct investment interactions and their effects. Asia-Pacific research and training network on trade, working paper series no. 45. Available at http://www.unescap.org/tid/artnet/pub/wp4507.pdf

Coe DT, Helpman E (1995) International R&D spillovers. Eur Econ Rev 39:859–887

Cole MA, Elliot RJ (2003) Determining the trade-environment composition effect: the role of capital, labor and environmental regulations. J Environ Econ Manage 46(3):363–383

Common M, Perrings C (1992) Towards an ecological economics of sustainability. Ecol Econ 6:7–34

Dasgupta S, Mody A, Roy S, Wheeler D (2001) Environmental regulation and development: a cross-country empirical analysis. Oxford Dev Stud 29(2):173–187

Dessus S, Bussolo M (1996) Politiques de l'environment et libéralisation des échanges au Costa Rica: une vue d'ensemble. OECD Development Center. Technical working paper no. 110, Paris, France

Esty DC, Gentry BS (1997) Foreign direct investment, globalization and environment. In: OECD, globalization and environment: preliminary perspectives. OECD, Paris, France

Gassebner M, Lamla MJ, Sturm J (2011) Determinants of pollution: what do we really know? Oxford Econ Papers 63:568–595

Gentry B (ed) (1998) Private capital flows and the environment. Lessons from Latin America. Edward Elgar Publishing, Cheltenham

Gentry B (1999) Foreign direct investment: boon or bane? In: Proceedings of OECD, foreign direct investment and the environment. OECD, Paris

Greening LA, Greene DL, Difiglio C (2000) Energy efficiency and consumption: the rebound effect. Energy Policy 28(6–7):389–401

Grossman GM, Krueger AB (1993a) Environmental impacts of the North American free trade agreement. In: Garber PM (ed) The Mexico-US free trade agreement. MIT Press, Cambridge

Grossman GM, Krueger AB (1993b) Pollution and growth: what do we know? In: Goldin I, Winters L (eds) The economics of sustainable development. MIT Press, Cambridge

Grossman GM, Krueger AB (1995) Economic growth and the environment. Quart J Econ 110(2):353–377

Guoming X, Cheng Z, Yangui Z, Shunqi G, Zhan JX (1999) The interface between foreign direct investment and the environment: the case of China. Report as part of the UNCTAD/DICM project, cross border environmental management in transnational economies, occasional paper no. 3, Copenhagen Business School

He J (2006) Pollution haven hypothesis and environmental impact of foreign direct investment: the case of industrial emission of sulfur dioxide (SO_2) in Chinese provinces. Ecol Econ 60:228–245

Hejazi W, Safarian E (2001) The complementarity between U.S. foreign direct investment stock and trade. Atlantic Econ J 29(4):420–437

Jackson T (2009) Prosperity without growth. Economics for a finite planet. Earthscan, London

Jha V (1999) Investment liberalization and environmental protection: conflicts and compatibilities in the case of India. Report as part of the UNCTAD/DICM project, cross border environmental management in transnational economies, Occasional paper, copenhagen business school

Liang FH (2006) Does foreign direct investment harm the host country's environment? Evidence from China. Working paper of the Haas School of Business, University of California, Berkeley, USA

Meers R (2000) A Test of the environmental Kuznets curve for local and global pollutants, Honors projects. Paper no. 76 of the Economic Department of the Illinois Wesleyan University

Munasinghe M (1999) Is environmental degradation an inevitable consequence of economic growth? Tunnelling through the environmental Kuznets curve. Ecol Econ 29:89–109

Murray CK (2013) What if consumers decide to all "go green"? Environmental rebound effects from consumption decisions. Energy Policy 54:240–256

Nguyen CD, Simpson G, Saal D, Nguyen AN, Pham NQ (2008) FDI horizontal and vertical effects on local firm technical efficiency. DEPOCEN working paper no. 17, pp 1–22

Nordström H, Vaughan S (1999) Trade and environment, vol 4. World Trade Organization, Geneva (Special studies)

O'Connor D (2000) Global capital flows and the environment in the 21st century. Organization for the economic co-operation and development. Technical paper no. 161. Paris, France
OECD (1997) Foreign direct investment and the environment: an overview of the literature. In: Document of the negotiating group on the multilateral agreement on investment (MAI). OECD, Paris, France
OECD (2001) Environmental outlook. Organization for Economic Co-operation and Development. Paris
OECD (2002b) Environmental benefits of foreign direct investment: a literature review. In: Document of the working party on global and structural policy. OECD, Paris, France
Opschoor JB (1995) Ecospace and the fall and rise of throughput intensity. Ecol Econ 15:137–140
Panayotou T (1997) Demystifying the environmental Kuznets curve: turning a black box into a policy tool. Environ Dev Econ 2(4):465–484
Panayotou T (2000) Globalization and the environment. Center for international development working paper no. 53, Harvard University, Boston
Perman R, Stern DI (2003) Evidence from panel unit root and cointegration tests that the environmental Kuznets curve does not exist. Aust J Agric Resour Econ 47(3):325–347
Rasiah R (1999) Transnational Corporations and the Environment: the case of Malaysia. Report as part of the UNCTAD/CBS Project cross border environmental management in transnational economies. Occasional paper no. 4, Copenhagen Business School
Selden TM, Song D (1994) Environmental quality and development: is there a Kuznets curve for air pollution emissions? J Environ Econ Manage 27:147–162
Shafik N (1994) Economic development and environmental quality. Oxford Econ Papers 46:757–773
Shafik N, Bandyopadhyay S (1992) Economic growth and environmental quality. Time series and cross-country evidence. Policy research working paper no. 904, World Bank, Washington, D.C.
Shahbaz M, Nasreen S, Afza T (2011) Environmental consequences of economic growth and foreign direct investment: evidence from panel data analysis. MPRA paper no. 32547
Sinani E, Meyer KE (2004) Spillovers of technology transfer from FDI: the case of Estonia. J Comp Econ 32(3):445–466
Stern DI (2004a) Environmental Kuznets curve. In: Cleveland CJ (ed) Encyclopedia of energy. Elsevier Inc, Amsterdam
Stern DI (2004b) The rise and fall of the environmental Kuznets curve. World Dev 32(8):1419–1439
UNCTAD (1999) World Investment Report 1999. United Nations Conference on Trade and Development, Geneva
UNCTAD (2007) World Investment Report 2007. United Nations Conference on Trade and Development, Geneva
WWF (2001) Preliminary assessment of the environmental and social effects of liberalization of tourism services. WWF international discussion paper, May, Geneva, Switzerland
Yandle B, Vijayaraghavan M, Bhattarai M (2002) The environmental Kuznets curve: a primer. Property and environment research centre (PERC) study no. 02-01. Available http://www.perc.org/articles/article688.php

Chapter 4
The Competition for FDI and Its Effect on Environmental Standards

4.1 Basic Features and Its Hypotheses

The competition for FDI and its effect on environmental standards represents the second vein of the discussion framework of the FDI-environment relationship and basically refers to the body of literature in which much of the debate on the FDI-environmental relationship has been developed so far.[1] It considers the development of a reflection process, whose mainstreams could be seen as two sides of the same coin: the impact of environmental standards on the location of firms' investment decisions and the environmental effects of international countries' competition for FDI. The first basically tries to understand if the existence of countries with different environmental regulations and standards can be a reason for firms relocating their activity.[2] The latter analyses the implication of the FDI-environment nexus which occurs, for example, when countries intentionally modify their environmental regulatory systems by lowering environmental standards to attract more FDI or—as a counter fact—by increasing them to gain a competitive advantage in the longer term. This body of literature deals with various phenomena associated with theories, whose existence is based on the existence of the following hypothesis: (1) "pollution havens"; (2) "race to the bottom"; (3) "regulatory chill".

---

[1] This is particularly true with regard to the issues of the location of firms' investment decisions as related to the "pollution haven" and "pollution halos" hypothesis, which will be explained later in this section.

[2] For the purpose of completion, it is worth referring that the aspect related to the environmental regulation is often seen to play a very minimal role in the decision process of investment location. In fact, in this body of literature, most of the works focus on various different aspects, such as the size, growth and accessibility of potential markets, political stability, labour costs, financial factors, mainly represented by the easy profits repatriation, red tape (administrative and legal) transparency and certainty, infrastructures, quality of life and education levels, etc. (Motta and Thisse 1994).

P. Pazienza, *The Relationship Between FDI and the Natural Environment*, SpringerBriefs in Economics, DOI: 10.1007/978-3-319-04301-2_4,

4.1.1 The "Pollution Haven" Hypothesis

The "pollution haven" is the hypothesis which is thought to occur when investors relocate their industries in those countries characterized by weaker or even absent environmental regulation, thus gaining the maximum advantage from producing at the lowest cost in light of environmental regulatory requirements. The search for pollution havens has widely characterized the debate and the literature on the FDI-environment relationship. Still nowadays the debate on this topic is lively. In fact, the production of scientific works is still unable to empirically show systematic evidence of the existence of pollution havens, while reaching contradictory results.

Although limited to a small number of cases and specifically considered sectors, some works prove the existence of the pollution havens hypothesis (Gray 2002).[3] One of the earliest works on this issue refers to an investigation to assess the relationship between the location of heavy-polluting industries in the United States and the dynamic of trade and investment data. No evidence confirming the existence of firms' behavioural patterns moving their investments to pollution havens in less developed countries was found in the study. This evidence was confirmed even for the case of the mineral processing sector, whose average FDI flow was much higher in developed than in developing countries (OECD 1997). Following the same reasoning, an updated reiteration of this work was carried out by a successive study and reached the same conclusion. With specific regard to the chemical industry of three Canadian provinces, the performed regression analysis did not find any significantly evident link between environmental regulation and plant location (OECD 1997). In 1990 a study focusing on how regional differences in environmental regulation can affect the car industry location decision did not reach any significant evidence with the exception of those countries characterized by heavy incompliance with air quality standards (Gray 2002). Another survey of more or less the same period found that 26 % of Maquiladora operators in Mexicali cited Mexico's lax environmental enforcement as an important reason for their location

[3] As made clear by some authors, the reason why a firm behaves in such a way can be better explained if we refer to the difference between the various types of FDI, which are normally divided into three types: (1) resource-seeking FDI; (2) production-seeking FDI; (3) market-seeking FDI. The first FDI category refers to those investors who aim to access critical primary resources which are not promptly available in their domestic market because of their physical scarcity or they are sold at much higher prices. For these kind of investment activities, final outputs are rather undifferentiated. As a result, a small difference in prices can mean larger market quotas. For this reason, this type of investment flow can be very sensitive to differences in environmental costs. The second category refers to those investments made abroad in export markets to provide platforms for production and sales. A typical example can be found in the car sector when observing the way in which car producers of one country expand their network in other countries. This type of investment is not very sensitive to an increase of environmental costs. The same can be said for the last category of investment, whose motivation is basically the need for investors to seek new opportunities to sell in larger markets, and even in those abroad (Esty and Gentry 1997).

there (WWF 2001). A milestone often recalled in the literature refers to an analysis of the United States Direct Investment Abroad (USDIA) data in 1992. In this work, it is observed how in the considered year developing and transitional economies received 45 % of the total flow. However, a very small quota of this flow (5 %) went to environmentally sensitive industries, such as those related to petroleum and gas, primary or fabricated metals, and chemical sectors, while a more significant proportion (24 %) reached already developed countries with tighter environmental standards. Hence, the conclusion supported the non-existence of the evidence that advanced countries export their "dirty" industries to less developed economies (Repetto 1995). As a counter fact, another study of an American public institution reported that a number of manufacturers in the wood furniture industry moved from the region of Los Angeles to Mexico between 1988 and 1990, because here they could use their solvents without considering any air pollution constraint (WWF 2001). Another study reaches a similar conclusion by using a statistical test to measure the effect of tighter environmental regulations on financial capital movement. In analysing the FDI outflow from various high and less-polluting US industries (chemicals, primary metals, electrical machinery, non-electrical machinery, food products and transportation equipments) to seven developing and 15 developed countries between 1985 and 1990, the study finds a significantly higher and positive correlation between those host countries with a more lenient environmental regulation and the US outflow of FDI. This evidence is particularly observed with regard to the chemical and primary metals sectors, which supports the existence of the pollution havens hypothesis for highly pollutant industrial sectors (Xing and Kolstad 2002). Similar evidence comes from other studies focusing on the Chinese investment context. One of them observes how a relevant number of highly pollutant foreign firms, dealing in the pesticides and asbestos production sectors, relocated their plants to China (Guoming et al. 1999). The other study refers how over 36 % of the Chinese inflow of FDI arrived to high-polluting production sectors such as, for example, printing, dyeing and electroplating (Gray 2002). Further examples of investment flight regarding industries involved in heavy-polluting production sectors are reported in another study. In the late eighties, firms in the wet processing and tanning industry relocated their activities from Europe to Brazil, as well as a number of firms working in asbestos tiles and benzedrine dye manufacturing facilities, which relocated to Mexico and Romania (Gray 2002). The analysis of more recent literature seems to be more supportive of the thesis on the existence of pollution havens. By econometrically evaluating the impact of the environmental stringency on the FDI outflow of OECD countries, a significant positive correlation is found. This evidence would support the pollution haven hypothesis as related to the industrial flight dynamic, which indeed corresponds to an increase of FDI outflows when environmental stringency of countries arises (Mihci et al. 2005). Among his various analysis focuses and conclusions, in his already-mentioned study, He (2006) provides convincing evidence of the existence of the pollution haven hypothesis. As has already been referred, he observes that the location and composition of the inward stock of Chinese FDI are highly motivated by pursuing a "production platform" with lower compliance costs of pollution regulation. A further study proves the existence of the hypothesis in question, while analysing the pollution abatement cost savings and FDI inflows to specific "dirty" production sectors in China (Wenhuda 2007). In the same direction, another study observes how firms in industries with higher abatement costs tend to invest more abroad to avoid high environmental compliance costs (Spatareanu 2007).

As counterfactual evidence in a very recent analysis carried out to understand whether or not ASEAN countries can be considered pollution havens for Japanese high-polluting industries, Elliot and Shimamoto (2008) provide indication of the non-existence of the pollution havens hypothesis. In addressing the question of why the literature fails to find more evidence of the "pollution havens" hypothesis, another study suggests that the lack of a systematic and firm link between industry abatement costs and the FDI outflow from developed countries is due to the fact that most of the studies ignore the role of factor endowments in the decision of MNCs to relocate their activities abroad.[4] As is commented in the study, by focusing particularly on the link between capital intensity and pollution intensity, it is possible to identify those countries which are more likely to be considered as pollution havens. Hence, after demonstrating the relationship between capital intensity and pollution intensity of US industries and the link between the stringency of countries' environmental regulations and capital abundance, the study econometrically analyses the determinants of the US multi-sector FDI outflow to Mexico and Brazil and finds the capital sectorial requirement a key determinant for FDI location. It also finds that in US industry the abatement cost of pollution levels is a significant determinant of its FDI outflow, thus proving the evidence of a pollution haven effect (Cole and Elliot 2005).

4.1.2 The "Race to the Bottom" Hypothesis

The race to the bottom hypothesis can be considered as a subset of the "pollution haven" phenomenon. In fact, it can be perceived as a possible factual evolution resulting from the recognition that "pollution havens" do matter. Indeed, it is natural to think that if the pollution haven hypothesis exists, then countries might

[4] As is commented, foreign investment, particularly that from North to South, is at least partially driven by factors endowment. Hence, it is plausible to expect that firms operating in capital intensive sectors—typically corresponding to pollution intensive industries—would invest in capital abundant countries—normally corresponding to those with stricter environmental regulations—whilst firms involved in labour intensive sectors—typically less pollutant—would invest in labour abundant countries characterized by a lenient or absent environmental regulations. This "Capital-Labour" hypothesis (KLH) seems to generate countervailing forces and dynamics with respect to those referred by the "pollution havens" hypothesis, which may explain why the literature testing the latter hypothesis fails to find systematic proof of its existence.

feel that by lowering their environmental standards they would result more competitive in FDI attraction. In fact, the race to the bottom phenomenon happens when, for example, a country's government undertake positive actions to lower its environmental standards with the final aim of bringing in FDI. Although the occurrence of this theoretical prescription may be plausible, little evidence is found to support its systematic validity.

In fact, some empirical evidence counters its theoretical foundation by highlighting that it seems unlikely that countries purposely proceed to lower their environmental standards, thus behaving in contrast to their own interest (Revesz 1992). In addition, the existence of some factors playing the role of countervailing forces in the race to the bottom should be considered. These might be basically related to the pressure arising from local communities, whose reasoning may follow the Not in My Backyard (NIMBY) principle (OECD 2002) and a number of other varying factors among which education and income levels can be seen among the most relevant aspects (Zarsky 1999: 60). Another study empirically testing the race to the bottom hypothesis focuses on the trends of air quality—measured in terms of suspended particulate matter—in the United States and in the three largest recipient countries of FDI in the developing world (Brazil, China and Mexico). The result shows how the globalization era has brought about a decline of the considered pollutant in major cities of all analysed countries, thus contradicting the theoretical foundation of the "race to the bottom" hypothesis. As is said, the lack of evidence is due to the fact that the basic assumption of the hypothesis in question misrepresents the political economy of pollution control in developing countries, because it does not consider a more realistic approach according to which low-income societies serve their own long-run interest by abating pollution (Wheeler 2001).

However, counterfactual evidence confirming the existence of the "race to the bottom" hypothesis can be observed especially with regard to case studies from specific sectors. The natural resources sector could be the fact in case, especially in developing countries, where the regulatory experience may be very limited and a preference for foreign investment is often shown. In Zimbabwe, for example, the dominant presence of foreign investors in the mining sector is explained by the national "Mines and Mineral Act" which takes over any other regulation including those related to norms of environmental protection (Gray 2002). Similar situations can also be observed in Indonesia and Papa New Guinea where, especially in the mining sector, governments have considerably relaxed environmental controls over mining operations in a range of areas. As is referred, in these two countries all mining operations are run under special conditions which require minimal or no regulation thus permitting an extensive detrimental effect on the environment. More precisely, in Indonesia mining corporations operate under special Contracts of Work (COW), thus being exempt from respecting environmental laws. Yet in Papua New Guinea, Indonesia and the Philippines, it is observed how governments have provided general or specific (project-by-project) exemptions from existing environmental and other laws with the aim of attracting higher flows of FDI (Mabey and McNally 1998: 39). Some further evidence in this direction is provided by other observations of the Canadian and German cases, where governments have simplified their environmental regulation by relaxing its enforcement, and implementing a more business-friendly context for investors (Esty and Geradin 1998). Another more relevant analysis supporting the existence of the hypothesis in question refers to the Costa Rica case study, where the government actively pursued investment projects in particular polluting sectors by skipping legal requirements, also including environmental aspects. This was made by attracting FDI through the Free Zone Law, whose implementation made the number of firms operating in the free zone context jump from 11 in 1986 to 183 in 1995 (Gentry 1998).[5]

4.1.3 The "Regulatory Chill"

The "regulatory chill" can be very briefly seen as a concept very closely related to the "race to the bottom" phenomenon. It occurs when countries desist from setting stricter environmental regulations and standards with the aim of avoiding loss of competitiveness in attracting FDI (Gray 2002).[6] As a consequence, environmental regulation can get "stuck in the mud". More specifically, behind this idea there is the conviction that more stringent pollution control requirements impose costs which will harm the competitiveness of the eventual regulated industry. This leads economic agents—especially industrialists—to argue for lower environmental standards than would be justifiable in the absence of the problem of competitiveness. This would generate a context characterized by a lack of political consensus and support (i.e. "the political drag" effect), which plays its relevant role in the "stuck at the bottom" process and confirms the idea expressed in various models where political institutions are seen as crucial variables of this problem (Porter 1999; Revesez 1992). In this sense, the evidence of a recent study is supported, which analyses the relationship between environmental regulatory systems and FDI from a completely different perspective. It develops a political economy model with imperfect product market competition and where domestic and foreign firms jointly lobby the local government for the introduction of a

[5] The work points out that the Free Zone Law did not set clear environmental requirements for companies entering Costa Rica and that the normative framework was confusing and incomplete. Furthermore, people working for governmental agencies involved in FDI attraction were unaware of environmental laws and norms. As a result, only two out of the 183 companies arriving in Costa Rica during that period operated under a formal environmental programme.

[6] In the case of developing countries, which are very often characterized by the inexistence of environmental regulations, this phenomenon is better known as the "stuck at the bottom" effect to mean that because of the fear of losing competitiveness in FDI attraction, they would remain "stuck at the bottom" with minimum or no environmental regulations. For example, during the last decade the governments of Morocco and Tunisia showed their unwillingness to upgrade the regulation level of the phosphate industry for fear that the companies operating in the sector would leave their countries and relocate the activities in places with lower environmental stringency (Gray 2002; Vogel 1995).

pollution tax while also considering the corruptibility degree of the investment host countries' governments. The empirical test, a panel data for 33 countries, finds that FDI affects environmental regulation depending on the governments' corruptibility level. If the corruptibility degree is high, FDI leads to a less stringent environmental regulation and vice versa (Cole et al. 2006). However, as is often highlighted, the regulatory chill refers to a scenario, which is very difficult to prove since it generally refers to government inaction in undertaking appropriate environmentally protective measures in response to FDI pressure. This is indeed something very difficult to demonstrate. For this reason, this specific issue is not characterized by the existence of wide research. Nevertheless the chilling effect on regulation seems to be the most relevant effect of policy competition between and within countries rather than the race to the bottom because of the reasons explained above. Although unproven through statistical analysis, some examples of the regulatory chill phenomenon (also occurring via a political drag effect) exist and are mainly related to cases where the competition for FDI or their holding within a country has been quoted as an important reason for not introducing new environmental regulations and taxes. For example, a number of large and important corporations operating in the oil sector threatened to reduce their investments in the Netherlands as soon as the government announced a plan to introduce a carbon tax. Similarly, in response to the possibility of introducing a carbon tax in UK, the paper federation announced that in such a case its associated industries would seriously consider relocating their activities to other countries, thus leaving the UK and all the European area (Mabey and McNally 1998: 40–41). Evidence supporting the opposite situation can also be found in the literature. For example, in Mexico, following an implementation phase of environmental regulation and enforcement, a survey-based analysis on the environmental management of manufacturing firms—mostly large US companies—found they were undertaking significant investments to meet the regulation requirements and inspections. The result of this study—which considering some views in literature we may see as a sort of edge between the "regulatory chilling" and the "race to the top" or "pollution halos" theories[7]—shows how less developed economies can tighten their environmental regulations without the fear that foreign investors will flee to other countries (Gentry 1998).

[7] However, it is worth highlighting that this actually clashes with the definition of the "race to the top" or "pollution halos" and seems to be more the result of a misunderstanding due to the fact that sometimes the concept of the "race to the top" theory comes out a little confused. In fact, in some works it is possible to read that the hypothesis in question occurs because a tighter environmental regulation stimulates technological advances, which promote efficiency and in turn attract investors (Gray 2002: 309). However, this idea should be more properly intended as the countervailing prospect of the main expectation related to the "regulatory chilling" theory, that is the existence of a weaker environmental regulation is more attractive for foreign investors. As will be highlighted, the "race to the top" or "pollution halos" hypotheses have nothing to do with the location decision of foreign investments, but with the environmental performance of foreign investors in the host country.

4.2 Some Concluding Considerations

As can be appreciated from the discussion above, the main point of this part of the literature on the competition for FDI and its effect on environmental standards is basically represented by the environmental cost of firms' production. Indeed, through all the hypotheses analysed the redundant idea is that to avoid higher environmental compliance costs, firms relocate their production to other countries where they can take advantage either of less stringent environmental regulations—implied by the "pollution havens" and the "race to the bottom" hypotheses—or of minor costs arising from the virtuous circle related to the theoretical foundation of the "pollution halos" hypothesis, that is a cost saving as a result of the existence of higher levels of technological efficiency. Hence, in this perspective the leniency or the stringency of the environmental regulation affects the firms' production costs and, as a result, can be thought to be determinative for the location of their investment decision. However, the literature very often highlights how environmental costs do not seem to be a strong motivating factor—like those mentioned in footnote 3 of this Chapter—in determining the location of foreign investment. According to what is generally stated in official documents and relevant studies, environmental costs represent a very small proportion of total costs if compared to labour and capital costs, so that any difference in environmental regulation will have little impact on a firm's location decisions (i.e. Dasgupta et al. 2001; WTO 1998; Motta and Thisse 1994).[8] However, as is generally claimed, research associated to this body of literature is still lacking a deeper comprehension of how environmental regulation plays a relevant role in the location of firms' investment decisions (Gray 2002). Although some evidence confirming the theoretical thinking associated to the considered hypotheses are found both at macro and micro-level analyses, a lack of its systematic proof still risks to leave the debate at an ideological dispute among policy decision makers.

References

Cole MA, Elliot RJ (2005) FDI and the capital intensity of 'dirty' sectors: a missing piece of the pollution haven puzzle. Rev Dev Econ 9(4):530–548

Cole MA, Elliot RJ, Fredriksson PG (2006) Endogenous pollution havens: does fdi influence environmental regulation? Scand J Econ 108(1):157–178

---

[8] Although this is not generally confirmed, particularly for specific sectors—as observed in the discussion above—it happens because countries commonly fail to properly price their environmental assets. According to what is broadly said in the context of environmental and natural resource economic studies, if external environmental costs are actually computed and internalized, then compliance costs would increase considerably, thus making environmental regulation a more relevant factor in the decision process of firms' investment location.

Dasgupta S, Mody A, Roy S, Wheeler D (2001) Environmental regulation and development: a cross-country empirical analysis. Oxf Dev Stud 29(2):173–187
Elliot RJR, Shimamoto K (2008) Are ASEAN countries havens for Japanese pollution-intensive industry? World Econ 31(2):236–254
Esty DC, Gentry BS (1997) Foreign direct investment, globalization and environment. Globalization and environment: preliminary perspectives. OECD, Paris
Esty DC, Geradin D (1998) Environmental protection and international competitiveness. J World Trade 32(3):5–47
Gentry B (ed) (1998) Private capital flows and the environment. Lessons from Latin America. Edward Elgar Publishing, Cheltenham
Gray KR (2002) Foreign direct investment and environmental impacts—is the debate over? Rev Eur Commun Int Environ Law 11(3):306–313
Guoming X, Cheng Z, Yangui Z, Shunqi G, Zhan JX (1999) The interface between foreign direct investment and the environment: the case of china. Report as part of the UNCTAD/DICM project, cross border environmental management in transnational economies, occasional paper no. 3, Copenhagen Business School
He J (2006) Pollution haven hypothesis and environmental impact of foreign direct investment: the case of industrial emission of sulfur dioxide (SO_2) in Chinese provinces. Ecol Econ 60:228–245
Mabey N, McNally R (1998) Foreign direct investment and the environment. From pollution havens to sustainable development. WWF-UK report
Mihci H, Cagatay S, Koska O (2005) The impact of environmental stringency of the foreign direct investments of the OECD countries. J Environ Assess Policy Manage 7(4):679–704
Motta M, Thisse JF (1994) Does environmental dumping lead to delocation? Eur Econ Rev 38:563–576
OECD (1997) Foreign direct investment and the environment: an overview of the literature. Document of the negotiating group on the multilateral agreement on investment (MAI), OECD, Paris
OECD (2002) Environmental issues in policy-based competition for investment: a literature review. Document of the working party on global and structural policy, OECD, Paris
Porter G (1999) Trade competition and pollution standards: 'race to the bottom' or 'stuck at the bottom'? J Environ Dev 8(2):133–151
Repetto R (1995) Jobs, competitiveness and environmental regulation: what are the real issues? World Resources Institute, Washington D.C
Revesz RL (1992) Rehabilitating interstate competition: rethinking the 'race to the bottom' rationale for federal environmental regulation. NY Univ Law Rev 67:1210–1254
Spatareanu M (2007) Searching for pollution havens: the impact of environmental regulations on foreign direct investment. J Environ Dev 16(2):161–182
Vogel D (1995) Trading up: consumer and environmental regulation in a global economy. Harvard University Press, Cambridge, MA
Wenhua D (2007) Pollution abatement cost savings and FDI inflows to polluting sectors in china. Environ Dev Econ 12(6):775–798
Wheeler D (2001) Racing to the bottom? Foreign investment and air pollution in developing countries. J Environ Dev 10(3):225–246
WTO (1998) World trade organisation annual report. World Trade Organization, Geneva
WWF (2001) Preliminary assessment of the environmental and social effects of liberalization of tourism services. WWF International Discussion Paper, Geneva, May
Xing Y, Kolstad CD (2002) Do lax environmental regulations attract foreign investment? Environ Resour Econ 21:1–22
Zarsky L (1999) Havens, halos and spaghetti: untangling the evidence about foreign direct investment and the environment. In: Proceedings of OECD foreign direct investment and the environment conference, OECD, Paris

[illegible]

[illegible]

[illegible]

[illegible]

[illegible]

[illegible]

[illegible]

[illegible] (2013) [illegible] and [illegible] environmental [illegible]

[illegible]

[illegible] The [illegible] and [illegible] OECD [illegible]

[illegible] (2008) [illegible] Trade [illegible] and the [illegible] [illegible] (2011) [illegible] of the [illegible]

[illegible]

Chapter 5
The Cross-Border Environmental Performance

5.1 Basic Features and Its Theoretical Foundation

The third and final considered cluster of research on cross-border environmental performance of foreign investing firms goes well beyond the understating of how FDI location can affect or can be affected by environmental regulation. According to some reflections, the particular consideration of the chosen environmental management approach by firms investing abroad appears to be relevant to the understanding of the environmental implication of FDI in host economies (Hansen 1999; UNCTAD 1999).[1] For this reason, much of the empirical analysis developed within this cluster of research focuses more on investigating the driving forces of the different environmental behaviours and performance of foreign firms in their cross-border and relatively undifferentiated operations. In other words, it takes into account the aspect of corporate environmental performance in an attempt to understand the reason why a certain type of environmental management approach is chosen by a foreign investing firm once it has established its activity abroad in a host country.[2]

[1] As is claimed, this still appears to be a largely unexplored field. For this reason a call for further research generally exists.

[2] In this context, for example, foreign investors can opt for a wide series of strategies from those aiming to achieve significant worldwide environmental results, such as those trying to internalize global environmental costs, to those which are followed only to accomplish local environmental regulations.

P. Pazienza, *The Relationship Between FDI and the Natural Environment*, SpringerBriefs in Economics, DOI: 10.1007/978-3-319-04301-2_5,

5.2 The "Pollution Halos" (or the "Race to the Top" Theory)

This part of the literature is relevantly characterized by the "pollution halos" or the "race to the top" theory.[3] This theory assumes exactly the opposite of the "pollution havens" and the "race to the bottom" hypotheses and has nothing to share with the location decision of the foreign investor. It basically focuses on how the investor performs from an environmental point of view once he gets into the host country, having based his investment there (OECD 2002a; Zarsky 1999). As has already been said, while the latter two hypotheses are based on the view that an inverse relationship exists between the strictness of environmental regulations and the location decision of foreign investments,[4] the "pollution halos" theory is based on a different point of view. Under the "Porter hypothesis", it could be said that stronger environmental policies would represent a comparative advantage for countries because, as a result of an over time process basically characterized by the spread of the positive implications of technological advances which promote innovation and efficiency, competitiveness (in terms of production cost reduction and improvement of product quality) would improve in the whole marketplace (Porter 1990). In clearer terms, the idea of the "pollution halos" hypothesis is based on the premise that the existence of political and economic pressures in a "greener" country might influence environmental practices in specific sectors of another less "green" country. Of course, the main expectation at the basis of this theory is that investing firms from developed areas perform environmentally better because they hold newer and cleaner technologies as well as better environmental management systems and best-practice. This can be the result of the stricter environmental regulations existing in their home countries or even the pressure from "greener" consumers at home for their products. On the basis of the assumption that MNCs apply the same production and management systems (also those related to the environmental aspect) regardless of the country in which they operate, then the expectation is that the FDI flow they move can represent a relevant vehicle for spreading technological advances and best-practice all over the world. This theory, however, is characterized by some weaknesses which can be synthesized in two

---

[3] The "pollution halos" phenomenon is also termed the "race to the top" phenomenon or the "California effect" (Gray 2002). The "California effect" takes its name from a specific case related to a virtuous circle which can be read in the history of American emissions standards. The Clean Air Act of 1970 gave to California—which chose this option—the possibility of undertaking stricter emission standards than those required for the rest of the US states. Later in 1990, Congress decided to bring national emission standards up to the Californian level and once again gave California the possibility of choosing a stricter emission level. At the same time, Congress also gave the other states the same option. In 1994, 12 states asked Congress to adopt the new standards set by California (Vogel 1995: 259).

[4] As has already been said, the main perception and expectation of these hypotheses is that a more stringent environmental regulation would increase production costs, reduce competition, thus making foreign investment go elsewhere.

main aspects. First of all, the literature fails to give systematic proof of its existence. Secondly, as has already been observed, the empirical existence of the “pollution halos” hypothesis—when verified—seems to be particularly related to the energy sector (Mabey and McNally 1998). Some studies support the existence of “pollution halos” in Mexico, Venezuela and Cote d’Ivoire where, through the use of the energy intensity levels as a proxy of the emissions from its consumption, it is found that foreign ownership firms are associated to lower levels of energy use thus resulting environmentally cleaner (Eskeland and Harrison 2003). Another study supporting the “pollution halos” hypothesis is related to an investigation of the investments in the field of electricity generation in China. Through the use of a questionnaire submitted to a number of specific respondents, it finds evidence that, thanks to advanced technologies, foreign investments are characterized by higher levels of efficiency in energy production and use—in absolute terms—and reduced emissions as a result (Blackman and Wu 1999). A number of other analyses, however, provide countervailing evidence. For example, three studies from different authors on Asian countries (Bangladesh, India and Indonesia) analysing fertilizers, pulping, paper and various other plants linked (financially or managerially) to OECD entities did not find any evidence of better environmental performance, this being associated to the scale of the plant (the bigger the cleaner) and particularly to the employment of newer technology, which is not necessarily associated to the FDI flow (Hettige et al. 1996; Huq and Wheeler 1993). In investigating the spending level on pollution abatement, another analysis already mentioned on the Korean manufacturing sector does not provide any evidence of the “pollution halos” hypothesis. Surprisingly, in fact, the result shows that domestic firms seem to perform better than foreign ones from an environmental perspective (Aden et al. 1999). To conclude, a further study on the Mexican manufacturing sector does not find any significant evidence that plants linked to OECD economies through MNCs investments, trade flows, management training or management experience put much effort—this measured in terms of adoption of ISO 14000 schemes and use of plant personnel for environmental control and inspections—into improving their environmental management strategies and performance. In addition, it does not find any significant evidence that plants with new technological equipment are cleaner and perform environmentally better (Dasgupta and Hettige 2000).

However, going beyond the detection of the studies supporting or not the “pollution halo” assumption, it is possible to observe the existence of works which identify the factors determining the environmental strategy choice of investing firms. Among the factors determining the environmental behaviour strategy, it is observed how foreign firms go from a centralized to a decentralized strategy approach depending on their size (UNCTAD 1999).[5] However, this is not the only

[5] As is reported, although this is not always the rule, larger MNCs keep decisions on the environmental strategy approach at their centralized level. In this sense, decisions are taken directly by the foreign investor and are applied despite the country it operates in. Instead, smaller MNCs decentralize these decisions to their foreign affiliate level.

determinant of the decision process, as other factors enter into the game. These can also be referred to another series of aspects such as ownership, market forces, industrial forces, and formal and informal regulatory forces. With regard to the ownership, the countervailing confirmation of the hypothesis that foreign-owned enterprises perform environmentally better than domestic ones is observable. In fact, some studies state that foreign investing firms are cleaner and more efficient in energy use and production, thus helping to reduce polluting emission levels (Eskeland and Harrison 2003). Other studies produce evidence of the better environmental performance of foreign-owned firms. For example, an investigation of MNCs working in various sectors of the Chinese economy shows how these generate a positive environmental effect through fulfilling environmental compliances and adopting environmental regulatory schemes such as ISO 14000 (Christmann and Taylor 2001). Different evidence is also observed in the literature, although this seems more alluded to the past. Some studies find no evidence to support the relationship between the ownership of firms and their environmental strategy and performance. This assertion is supported by a study on the reality of South and South-East Asia in which the ownership structure of firms is not found to be a significant determinant for pollution abatement (Hettige et al. 1996). The same evidence is highlighted in other studies focusing on Mexico, where firms' environmental strategy and performance are not found to be affected by the ownership structure, but other aspects such as plant size and public environmental awareness can generate a direct effect on the firms' environmental performance (Dasgupta and Hettige 2000). A completely different view is expressed in another analysis referring to the Korean manufacturing sector, in which it is observed that domestic firms perform better than foreign firms from an environmental point of view (Aden et al. 1999). The industrial forces argument focuses on the featuring aspects of an industry. It is argued that the environmental performance of an industry depends on features such as its concentration, collaboration or collusion behaviours existing in the market system, since this will affect environmental cross-border practices among its members. Better environmental performances are more likely to occur in the case of highly concentrated industry, such as the case of oligopoly, or when a collaborative approach is followed. In the first case, firms will find themselves in a better situation to compensate environmental costs because of their higher capacity to control the market. In the second case, by following a collaborative behavioural approach, firms have greater motivation to develop common codes of conduct and to implement best practices (Hansen 1999). A further study focuses on industrial forces by pointing out that one of the beneficial aspects of globalization is the encouragement firms receive in undertaking environmental self-regulation. This basically happens through the adoption of voluntary environmental management schemes, which often go much further than local government regulation requirements (Christmann and Taylor 2001). The issue of market forces refers to commercial and financial featuring factors of a marketplace which can induce firms and foreign investors to implement better strategies with the aim of achieving greater environmental performance. This situation can lead to a competitive advantage in terms of positive implications in the relationship between FDI and the environment. For a better explanation of this aspect, the literature often proposes two case studies. The first regards the case of an environmental certification scheme in the banana production sector of Costa Rica, where pressure from the consumers in foreign markets and the need to produce at a lower cost led the firms in the sector—mainly of foreign investors—to reduce the use of chemicals and irrigation, thus ensuring better environmental results. The second example refers to the Brazilian pulp and paper production sector. As in the previous case, this sector is observed to have improved its environmental performance by creating a national system of certification (Gentry 1998). The fact that customer pressure—as a featuring aspect of the marketplace—can push firms towards better environmental performance is also observed in China, where a significant quota of those producing and selling products to developed countries have higher environmental compliances and are more likely to adopt the ISO 14000 environmental schemes (Christmann and Taylor 2001). The other market feature can be seen in the financial aspect and is related to the dynamic of the stock performance. This is an aspect which can also influence firms in the same way as has already been commented. For instance, as proven in some studies, information on negative environmental performance in poor environments harmed capital markets of countries such as Argentina, Chile, Mexico and Philippines (Dasgupta et al. 2001). Similarly, a case study report on the Indonesian pulp and paper industry shows how the larger producer in the sector gained enormous benefits—in terms of a significant increase of the company's stock value—from the excellent environmental improvement achieved (World Bank 2000). More support to this view comes from a study which shows how American MNCs adopting higher environmental standards in their foreign operations than those required in domestic operations have a higher market value with respect to those not following the same strategy (OECD 2002b). The aspects of formal regulatory forces refer to the consideration that investing companies abroad are normally influenced by the host country regulations (O'Connor 2000). A study already mentioned in this work, for example, observes that in Mexico, due to the existence of a good and well enforced environmental regulation, a number of American investors in its manufacturing sector generated positive technological effects, especially by bringing advanced water treatment facilities, thus ensuring a higher level of eco-efficiency in their business (Gentry 1998). However, this observation is not always confirmed. In fact, it should also be considered that the implementation and enforcement of legal rules, especially in the environmental issue, can be particularly problematic with the result of generating very weak or null results. Evidence in this sense is particularly related to a deficiency in monitoring the respect of rules and is observed in various analyses. For example, weaknesses and enforcement problems are referred in China (Guoming et al. 1999), Malaysia (Rasiah 1999) and India (Jha 1999). The existence of these difficulties would require control power to be worked out at some levels. Following the literature, it is possible to note that this could potentially be ensured by investment source countries which can set out binding environmental rules for their corporation operating abroad. Although the refection has been made and brought to light by various UNCTAD reports, no empirical work seems to have been conducted in this area, which is said to suffer from a lack of research (OECD 2002b). However, when the regulatory forces of the formal sector (of both investment host and source countries) are unable to govern or to affect the environmental conduct of corporations, then it is easier that subjects of the informal regulatory forces make their appearance. As often happens, non-state actors—such as Non-Governmental Organizations (NGOs)—appear and try to cover the regulation gap by playing their role. This would be particularly aimed at implementing activities of so-called "civil regulation"—normally run in the investment source country but also in the host country—and at achieving the reduction of the environmental impact deriving from business activities through better public environmental information, that is by exposing public opinion to corporate behaviour and promoting the adherence of corporations to codes of responsible business conduct (Newell 2001).

5.3 Some Concluding Considerations

As is clear, this final body of literature considering the cross-border environmental performance of foreign firms actually focuses on what is also known as corporate environmental performance. In recent years, this has become a relevant public issue with respect to the situation of both developed and developing countries. Research does not yet properly explain why cross-country differences of MNC environmental performance is observed. As is generally referred, this can be due to the lack of appropriate data and more comprehensive knowledge linked to the fact that the release of information on the environmental aspects of the operation of foreign firms abroad still remains a voluntary fact. As the search for public consensus pushes firms to release environmental reports of their activities, more information will become available for additional investigation in this area of discussion. This will help to better understand the role and relevance that firms' features (as endogenous drivers) and market, industry and regulatory forces (as exogenous drivers) have on their environmental performance in both domestic and foreign countries.

References

Aden J, Kyu-hong A, Rock MT (1999) What is driving the pollution abatement expenditure behavior of manufacturing plants in Korea? World Dev 27(7):1203–1214

Blackman A, Wu X (1999) Foreign direct investment in China's power sector: trends, benefits and barriers. Energy Policy 27(12):695–712

Christmann P, Taylor G (2001) Globalization and the environment: determinants of self-regulation in China. J Int Bus Stud 32(3):439–458

Dasgupta S, Hettige H (2000) What improves environmental performances. Evidence from the Mexican industry. J Environ Econ Manag 39(1):39–67
Dasgupta S, Mody A, Roy S, Wheeler D (2001) Environmental regulation and development: a cross-country empirical analysis. Oxford Dev Stud 29(2):173–187
Eskeland GS, Harrison AE (2003) Moving to greener pastures? Multinationals and the pollution-haven hypothesis. J Dev Econ 70(1):1–24
Gentry B (ed) (1998) Private capital flows and the environment. Lessons from Latin America. Edward Elgar Publishing, Cheltenham
Gray KR (2002) Foreign direct investment and environmental impacts—is the debate over? Rev Eur Commun Int Environ Law 11(3):306–313
Guoming X, Cheng Z, Yangui Z, Shunqi G, Zhan JX (1999) The interface between foreign direct investment and the environment: the case of China, report as part of the UNCTAD/DICM project, cross border environmental management in transnational economies, occasional paper no. 3. Copenhagen Business School
Hansen MW (1999) Cross border environmental management in transnational corporations. An analytical framework, report as part of the UNCTAD/CBS project cross border environmental management in transnational economies, occasional paper no. 5. Copenhagen Business School
Hettige H, Huq M, Pargal S, Wheeler D (1996) Determinants of pollution abatement in developing countries: evidence from South and Southeast Asia. World Dev 24(12):1891–1904
Huq M, Wheeler D (1993) Pollution reduction without formal regulation: evidence from Bangladesh, World Bank working paper no 39. World Bank, Washington
Jha V (1999) Investment liberalization and environmental protection: conflicts and compatibilities in the case of India. Report as part of the UNCTAD/DICM project, cross border environmental management in transnational economies, Occasional paper, copenhagen business school
Mabey N, McNally R (1998) Foreign direct investment and the environment. From pollution havens to sustainable development, WWF-UK Report
Newell P (2001) Managing multinationals: the governance of investment for the environment. J Int Dev 13:907–919
O'Connor D (2000) Global capital flows and the environment in the 21st century, organization for the economic co-operation and development, technical paper no. 161. Paris, France
OECD (2002a) Environmental issues in policy-based competition for investment: a literature review, document of the working party on global and structural policy. OECD, Paris, France
OECD (2002b) Environmental benefits of foreign direct investment: a literature review, document of the working party on global and structural policy. OECD, Paris, France
Porter ME (1990) The competitive advantage of nations. Free Press, New York
Rasiah R (1999) Transnational corporations and the environment: the case of Malaysia, report as part of the UNCTAD/CBS project cross border environmental management in transnational economies, occasional paper no. 4. Copenhagen Business School
UNCTAD (1999) World investment report 1999, United Nations conference on trade and development. Geneva, Switzerland
Vogel D (1995) Trading up: consumer and environmental regulation in a global economy. Harvard University Press, Cambridge
World Bank (2000) Greening industry. New roles for communities, markets and governments. World Bank and Oxford University Press, Washington
Zarsky L (1999) Havens, Halos and Spaghetti: Untangling the evidence about foreign direct investment and the environment. In: OECD, foreign direct investment and the environment, conference proceedings. OECD, Paris

Chapter 6
Recent Trends and Prospects in the FDI Dynamic

6.1 The Macro-Geographical Distribution of FDI

In this chapter we examine some recent trends and prospects in FDI by reporting the information dispatched by the UNCTAD. More details on the global and regional trends and prospects of the FDI dynamic can be found in its World Investment Report (WIR) series, which—at the time of writing—shows the WIR 2012 as the last available update. Grasping information from the UNCTAD statistical database—which is synthesized in Tables 6.1 and 6.2[1]—and the mentioned investment report, it is possible to observe how the global FDI inflow significantly increased between 2002 and 2007 moving from about 627,975 million US$ in 2002 to 1,975,537 million US$ in 2007. In 2008 and 2009, the flow decreased as a result of the global economic turmoil. However, despite the financial and economic crisis of these two years and the ongoing sovereign debt crisis, the global inflow of FDI increased by 16 % between 2010 and 2011. More specifically, it rose from about 1,309,001 million US$ to about 1,524,422 million US$ showing an ameliorated situation with respect to what could be observed at the pre-crisis average level recorded between 2005 and 2007.[2]

The FDI flow reaching developed countries also grew consistently during the entire considered time. It rose from 443,431 million US$ in 2002 to 1,310,425 in 2007. After a significant fall in the next two years, which brought the investment level to 606,212 million US$ in 2009, the FDI flow began to go up again to reach 618,586 million in 2010 and 747,860 million in 2011. This performance shows an increase of the investment inflow of 21 % between 2010 and 2011, although it is still a quarter below the level of the pre-crisis three-year average.

---

[1] As a result of the reason mentioned in paragraph 1.3, where the reporting of FDI in terms of stock has already been said to be unsatisfactory because it is expressed in terms of "book value", here we only focus our comment on the flow of FDI. However, for the purpose of completeness, we also report Tables 6.3 and 6.4 where the data on the stock of FDI are presented.

[2] For brevity we base our observation and comment only on the FDI flow dynamic. Nevertheless, we also report the statistic data related to the FDI stock for completeness. We do not proceed to comment on the FDI stock dynamic since it basically echoes that performed by the FDI flow.

P. Pazienza, *The Relationship Between FDI and the Natural Environment*, SpringerBriefs in Economics, DOI: 10.1007/978-3-319-04301-2_6,

Table 6.1 FDI inflow and outflow in considered regions in 2002–2011

Region/country	1981–2001*	2002	2003	2004	2005	2006	2007	2008	2009	2010	2011
Inward flows (in million US$ at current prices and current exchange rate)											
World	339,479.1	627,974.8	586,956.4	744,329.2	980,727.1	1,463,351.2	1,975,537.0	1,790,705.7	1,197,823.7	1,309,001.3	1,524,422.2
Developed economies	248,194.3	443,431.7	376,807.6	422,179.1	622,625.4	981,869.3	1,310,425.4	1,019,648.0	606,212.3	618,586.1	747,860.0
Developing economies	88,389.8	173,283.0	190,124.8	291,866.0	327,247.8	427,163.4	574,311.5	650,016.8	519,225.0	616,660.7	684,399.3
Transition economies	2,895.0	11,260.1	20,023.9	30,284.1	30,854.0	54,318.4	90,800.1	121,040.9	72,386.4	73,754.5	92,162.9
Total OECD	274,699.3	205,046.7	186,699.8	222,196.6	231,225.9	366,368.8	586,850.3	693,359.0	841,108.7	875,218.3	621,441.6
European union	140,845.7	312,003.1	274,292.2	225,900.9	499,375.5	585,030.2	853,965.6	542,242.4	356,631.5	318,277.4	420,715.2
Outward flows (in million US$ at current prices and current exchange rates)											
World	335,525.8	528,495.9	570,679.3	925,716.1	888,560.8	1,415,093.9	2,198,025.0	1,969,336.0	1,175,108.4	1,451,364.7	1,694,396.1
Developed economies	300,221.0	476,341.0	513,209.9	788,795.1	741,744.2	1,152,033.6	1,829,578.1	1,580,752.9	857,792.0	989,576.4	1,237,507.6
Developing economies	34,468.5	47,484.3	46,667.6	122,791.6	132,507.0	239,336.0	316,863.5	328,120.8	268,476.0	400,144.1	383,753.7
Transition economies	836.3	4,670.6	10,801.8	14,129.5	14,309.6	23,724.3	51,583.5	60,462.3	48,840.4	61,644.2	73,134.8
Total OECD	300,529.1	2,271,189.4	2,449,861.2	3,935,539.2	3,653,673.3	5,716,725.8	9,177,470.3	7,892,658.6	4,611,633.2	5,556,892.5	6,688,833.9
European Union	187,510.5	259,864.4	290,173.3	371,478.0	604,075.6	691,763.9	1,204,747.4	957,797.6	393,618.3	482,904.6	561,805.0

*Annual average
Source UNCTADstat

Table 6.2 FDI inflow and outflow in considered regions in 2002–2011 (% of the total)

Region/country	1981–2001*	2002	2003	2004	2005	2006	2007	2008	2009	2010	2011
Inward flows (in %)											
World	100.0	100.0	100.0	100.0	100.0	100.0	100.0	100.0	100.0	100.0	100.0
Developed economies	73.1	70.6	64.2	56.7	63.5	67.1	66.3	56.9	50.6	47.3	49.1
Developing economies	26.0	27.6	32.4	39.2	33.4	29.2	29.1	36.3	43.3	47.1	44.9
Transition economies	0.9	1.8	3.4	4.1	3.1	3.7	4.6	6.8	6.0	5.6	6.0
OECD	80.9	32.7	31.8	29.9	23.6	25.0	29.7	38.7	70.2	66.9	40.8
EU	41.5	49.7	46.7	30.3	50.9	40.0	43.2	30.3	29.8	24.3	27.6
Outward flows (in %)											
World	100.0	100.0	100.0	100.0	100.0	100.0	100.0	100.0	100.0	100.0	100.0
Developed economies	89.5	90.1	89.9	85.2	83.5	81.4	83.2	80.3	73.0	68.2	73.0
Developing economies	10.3	9.0	8.2	13.3	14.9	16.9	14.4	16.7	22.8	27.6	22.6
Transition economies	0.2	0.9	1.9	1.5	1.6	1.7	2.3	3.1	4.2	4.2	4.3
OECD	89.6	429.7	429.3	425.1	411.2	404.0	417.5	400.8	392.4	382.9	394.8
EU	55.9	49.2	50.8	40.1	68.0	48.9	54.8	48.6	33.5	33.3	33.2

*Annual average

Source our computations on UNCTADstat

Table 6.3 Inward and outward FDI stocks in considered regions in 2002–2011

Region/country	1981–2001*	2002	2003	2004	2005	2006	2007	2008	2009	2010	2011
Inward stocks (in million US$ at current prices and current exchange rates)											
World	2,962,593.3	7,501,217.3	9,387,520.6	11,100,663.7	11,563,007.5	14,300,408.9	17,901,073.1	15,451,284.8	18,041,009.2	19,906,661.7	20,438,198.7
Developed economies	2,190,298.0	5,654,946.7	7,253,576.8	8,577,849.8	8,577,412.4	10,550,382.4	12,738,241.9	10,812,730.3	12,296,706.1	12,890,908.5	13,055,902.8
Developing economies	757,971.8	1,730,851.8	1,979,882.6	2,325,396.9	2,712,819.7	3,355,031.0	4,487,488.8	4,214,287.2	5,120,181.9	6,256,066.3	6,625,031.7
Transition economies	14,323.5	115,418.8	154,061.3	197,417.1	272,775.4	394,995.5	675,342.3	424,267.3	624,121.3	759,686.9	757,264.2
Total OECD	2,292,618.6	6,065,296.8	7,688,400.5	9,070,541.7	9,178,225.7	11,204,992.5	13,501,177.8	11,474,842.2	13,214,516.3	14,054,770.1	14,261,941.2
European Union	958,524.9	2,958,991.7	3,923,359.7	4,800,899.6	4,731,893.0	5,983,519.7	7,503,020.3	6,653,978.9	7,322,963.0	7,289,628.8	7,275,621.6
Outward stocks (in million US$ at current prices and current exchange rates)											
World	3,055,446.4	7,785,795.5	9,916,512.4	11,694,926.6	12,464,846.8	15,697,204.3	19,272,590.8	16,342,808.9	19,325,745.6	20,864,846.1	21,168,488.7
Developed economies	2,756,696.0	6,811,666.8	8,823,480.2	10,413,691.9	10,951,816.6	13,636,336.3	16,367,069.9	13,648,378.2	16,152,431.7	17,144,627.8	17,055,963.5
Developing economies	293,269.2	909,003.0	998,158.7	1,169,747.3	1,360,935.3	1,837,998.5	2,517,784.7	2,463,068.9	2,834,914.5	3,313,807.7	3,705,410.3
Transition economies	5,481.3	65,125.7	94,873.4	111,487.3	152,094.9	222,869.6	387,736.2	231,361.9	338,399.5	406,410.6	407,114.8
Total OECD	2,748,624.2	6,913,179.4	8,918,357.6	10,528,618.9	11,095,453.3	13,807,628.0	16,599,613.7	13,952,748.7	16,643,731.0	17,699,315.3	17,726,906.8
European Union	1,198,778.6	3,716,478.0	4,826,124.1	5,551,909.1	5,742,144.2	7,216,880.2	8,738,232.0	8,115,395.9	9,127,113.8	9,243,523.3	9,198,831.8

*Annual average
Source UNCTADstat

Table 6.4 Inward and outward FDI stocks in considered regions in 2002–2011 (% of the total)

Region/country	1981–2001*	2002	2003	2004	2005	2006	2007	2008	2009	2010	2011
Inward stocks (in million US$ at current prices and current exchange rates)											
World	100.0	100.0	100.0	100.0	100.0	100.0	100.0	100.0	100.0	100.0	100.0
Developed economies	73.9	75.4	77.3	77.3	74.2	73.8	71.2	70.0	68.2	64.8	63.9
Developing economies	25.6	23.1	21.1	20.9	23.5	23.5	25.1	27.3	28.4	31.4	32.4
Transition economies	0.5	1.5	1.6	1.8	2.4	2.8	3.8	2.7	3.5	3.8	3.7
Total OECD	77.4	80.9	81.9	81.7	79.4	78.4	75.4	74.3	73.2	70.6	69.8
European union	32.4	39.4	41.8	43.2	40.9	41.8	41.9	43.1	40.6	36.6	35.6
Outward stocks (in million US$ at current prices and current exchange rates)											
World	100.0	100.0	100.0	100.0	100.0	100.0	100.0	100.0	100.0	100.0	100.0
Developed economies	73.9	75.4	77.3	77.3	74.2	73.8	71.2	70.0	68.2	64.8	63.9
Developing economies	25.6	23.1	21.1	20.9	23.5	23.5	25.1	27.3	28.4	31.4	32.4
Transition economies	0.5	1.5	1.6	1.8	2.4	2.8	3.8	2.7	3.5	3.8	3.7
Total OECD	77.4	80.9	81.9	81.7	79.4	78.4	75.4	74.3	73.2	70.6	69.8
European union	32.4	39.4	41.8	43.2	40.9	41.8	41.9	43.1	40.6	36.6	35.6

*Annual average

Source our computations on UNCTADstat

In addition, the investment flow to developing and transition economies increased between 2002 and 2008. On the one hand, developing countries experienced an increase from 173,283 million US$ in 2002 to about 650,017 million in 2008, which was followed by a decrease to 519,225 million in 2009 and a new increase to about 616,661 million and 684,399 in 2010 and 2011 respectively. On the other hand, the investment flow in transition economies jumped from about 11,260 million US$ in 2002 to about 121,041 million in 2008. A decrease in 2009, which brought the level to 72,386 million, was followed by two new increases to about 73,754.5 in 2010 and about 92,163 million US$ in 2011. The contribution of the FDI inflow in both developing and transition economies to the global flow of inward FDI constantly increased over the whole considered period. It moved from about 29 % in 2002 to about 53 % in 2010 and 51 % in 2011 (when their contribution was 45 and 6 % respectively). This growth was also recorded during the worst years of the global economic turmoil and shows the economic dynamism and the strong role they can play in the future. According to details given in the UNCTAD (2012) report, the increase of FDI to developing economies was generated by an increase of 10 % in Asia and 16 % in Latin America and the Caribbean.

We developed Table 6.1 with the aim of highlighting the role of the OECD and European Union areas in the distributional dynamic of the FDI inflow. As can be observed, the FDI inflow reaching the OECD countries[3] was about 205,047 million US$ in 2002 and—apart from a reduction to about 186,700 million in 2003—increased during all the considered period (even in the years of the economic crisis) arriving to a level of 875,218 million in 2010. A new decrease was recorded in 2011 when the amount fell to about 621,442 million US$. In terms of contribution to the global FDI inflow dynamic, it can be observed how in 2002 the OECD area attracted about 32 % of the total world flow. This percentage almost constantly declined over the next years to reach a level of about 30 % in 2007. In the years of the global economic crisis, it rose to 38.7 % in 2008 and 70.2 % in 2009 to fall down to about 67 % in 2010 and 41 % in 2011.

The European Union area is one of the major attractors of the FDI inflow worldwide. In 2002 it attracted 312,003 million US$ (that is 50 % of the FDI inflow moving worldwide). After a decline in the next two years, the amount rose to about 542,242 million in 2008 (about 30 % of the FDI global inflow). After a new decline in 2009 and 2010, when the amount reached 356,631 (about 30 % of the global inflow) and 318,277 (about 24 %) million US$ respectively, the FDI

[3] The 30 OECD countries we are considering here are: (1) Australia; (2) Austria; (3) Belgium; (4) Canada; (5) Czech Republic; (6) Denmark; (7) Finland; (8) France; (9) Germany; (10) Greece; (11) Hungary; (12) Iceland; (13) Ireland; (14) Italy; (15) Japan; (16) Korea Republic; (17) Luxembourg; (18) Mexico; (19) Netherlands; (20) New Zealand; (21) Norway; (22) Poland; (23) Portugal; (24) Slovak Republic; (25) Spain; (26) Sweden; (27) Switzerland; (28) Turkey; (29) United Kingdom; (30) United States of America. The remaining four OECD countries (Chile, Estonia, Israel and Slovenia) are not taken into consideration because their accession only took place in 2010.

inflow rose again to 420,715 million, although at a lower level than those recorded in the four pre-crisis years.

A very short analysis of the outward FDI makes us observe how its flow from developed countries continuously increased from 2002 to 2007 moving from 476,341 to about 1,830 billion US$. After a decrease recorded during the crisis period, it returned to a consistent increase of about 25 % between 2010 and 2011, reaching about 1.237 billion US$ in the last considered year. This increase was the result of the operations deriving from the three major developed-economy investor blocs (the European Union, North America and Japan), although the driving factors differed for each. FDI from the United States was driven by a record level of reinvested earnings (82 % of total FDI outflows), partially driven by TNCs building on their foreign cash holdings. The rise of FDI outflows from the European Union was determined by cross-border M&As. Japanese TNCs doubled their FDI outflow through M&A purchases in North America and Europe (+132 %) as a result of an appreciation of its national currency (UNCTAD 2012).

The outward FDI flow generated by developing economies shows a generally decreasing trend between 2002 and 2010. With more specific regard to the last two years of the considered period, it shows a decline of 4 % moving from 400,144 to about 383,754 % to $384 billion between 2010 and 2011, although its contribution to the global FDI outflow remained at 23 %. In the last considered year, the flows from Latin America and the Caribbean fell by 17 %. Among the main reasons for this decrease is the repatriation of capital to the considered country areas (counted as negative outflows), which was partially motivated by financial considerations (i.e. exchange rates, interest rate differentials). Once again, with reference to the last year of our considered period, the FDI outward flow from East and South–East Asia was largely stagnant. A 9 % decline was recorded with respect to the outflow of FDI from East Asia, while outward FDI from West Asia increased significantly to $25 billion (UNCTAD 2012).

6.2 The Sectoral Distribution of FDI

To comment on the sectoral distribution of the FDI flows, an analysis of the statistics between 2008 and 2011 allows us to observe how it rose in all three sectors of primary, manufacturing and services. According to FDI projects data (comprising cross-border M&As and greenfield investments) dispatched by UNCTAD and reported in Table 6.5 here, the FDI flow destined to projects in the primary sector rebounded in 2011 to a level of 200 billion US$ after a significant fall in 2009 and 2010. An analogous situation can be the investment dynamic in the service sector, which reversed the negative trend of the previous two years and reached 570 billion US$ in 2011. As can be appreciated by looking at the part of the table where the percentages are shown, the share of these two sectors rose slightly at the expense of manufacturing, which reached 660 billion US$ in 2011 after having experienced a sharp decline in the two previous years. From what is

Table 6.5 Sectoral distribution of projects related to the global FDI flow (2008–2011)

Year	Value (in billion US$)			%		
	Primary	Manufacturing	Service	Primary	Manufacturing	Service
2005–2007*	130	670	820	8	41	51
2008	230	980	1,130	10	42	48
2009	170	510	630	13	39	49
2010	140	620	490	11	50	39
2011	200	660	570	14	46	40

*Annual average
Source UNCTAD, world investment report (2012)

said in the last world investment report by UNCTAD (2012), it is possible to learn how the top five industries contributing to the rise in FDI projects were extractive industries (mining, quarrying and petroleum), chemicals, utilities (electricity, gas and water), transportation and communications, and other services (largely driven by oil and gas field services) (Table 6.5).

With regard to the prospects, it must be highlighted that because of the economic uncertainty and the probability of lower growth rates in major emerging markets there is an objective risk of seeing the generally favourable trend of the last couple of years undercut. UNCTAD foresaw the FDI growth rate would slow in 2012. Leading indicators are suggestive of this trend, with the value of both cross-border M&As and greenfield investments retreating during the considered year. As a result, it was believed highly likely that in 2012 the FDI inflow would grow very moderately in all the three identified macro-regions, namely developed, developing and transition economies. Among the countries belonging to the developing regions, Africa is referred as a particular case since its FDI inflow is expected to increase. A very moderate FDI growth is also expected to happen in Asia (including East and South–East Asia, South Asia and West Asia) and Latin America. Similarly to Africa, however, the FDI flows to transition economies was also expected to grow further during 2012 and in the next couple of years (UNCTAD 2012).

To conclude this section, we move very briefly onto referring to the prospects expected for the next few years. In the short-medium run UNCTAD predictions show that, although investor uncertainty is still high, the global FDI flow will continue to grow at a modest but stable pace reaching 1.8 trillion US$ in 2013 and 1.9 trillion US$ in 2014, barring any macroeconomic shock (UNCTAD 2012). This should be the result of the feeling recorded by the World Investment Prospects Survey (WIPS), which is run yearly by UNCTAD and pools the Chief Executive Officers (CEOs) of TNC on their investment plans. This yearly survey reveals that, although the number of pessimistic CEOs is 10 % higher than the number of those optimistic, the largest group of respondents (about 50 %) are neutral or undecided. However, a more optimistic view is expressed by respondents in the longer run after the end of 2012. In fact, more than half of them foresees an increase—compared with the situation in 2011—of their planned FDI expenditure between 2012 and 2014. Although the forecasted increase of the FDI

Table 6.6 Summary of econometric results of medium-term baseline scenarios of FDI flows by region (in billion US$)

Host region	Averages		FDI flows			Projections		
	2005–2007	2009–2011	2009	2010	2011	2012	2013	2014
Global FDI flows	**1,473**	**1,344**	**1,198**	**1,309**	**1,524**	**1,495–1,695**	**1,963–1,925**	**1,700–2,110**
Developed countries	**972**	**658**	**606**	**619**	**748**	**735–825**	**810–940**	**840–1,020**
European Union	646	365	357	318	421	410–450	430–510	440–550
North America	253	218	165	221	268	255–285	280–310	290–340
Developing countries	**443**	**607**	**519**	**617**	**684**	**670–760**	**720–855**	**755–930**
Africa	40	46	53	43	43	55–65	70–85	75–100
Latin America and the Caribbean	116	185	149	187	217	195–225	215–265	200–250
Asia	286	374	315	384	423	420–470	440–520	460–570
Transition economies	**59**	**79**	**72**	**74**	**92**	**90–110**	**100–130**	**110–150**

Source UNCTAD (2012)

flow will be driven by developed economies, indications also suggest that in the short-medium term developing and transition countries will continue to keep up with global FDI growth. Responses given by CEOs to the WIPS of 2012 rated six developing and transition economies among their 10 future investment destinations by the end of 2014, with Indonesia among the five top destinations for the first time (UNCTAD 2012). Table 6.6 reports the projection of FDI growth in the short-medium run, resulting from the econometric analysis performed for the WIR 2012.

6.3 Some Concluding Considerations

This chapter has presented the main trend and prospects of the FDI phenomenon. It has highlighted how the global flow of FDI will grow moderately in the short-medium term, although economic uncertainty levels still exist. The further FDI growth expected in the next few years represents a condition which should push us to make more effort in understanding whether and how FDI affects the environment with the aim of producing as much evidence as possible to contribute to a better understanding of the associated dynamics. In this way, it will be possible to help the policy-making activity of governments to pursue a more conscious production of regulations for the implementation of more sustainable ways of managing investment activities. With regard to this, it is useful to conclude by repeating a concept already expressed in the previous pages: whether FDI is really functional for development and, in particular, for sustainable development, depends especially on the way it is managed by the receiving countries' governments, which in turn depends on their vision of economic development and environmental conservation management. In other words, it all depends on their policy and regulatory frameworks.

References

UNCTAD (1993) World investment report 1993, United Nations conference on trade and development. Geneva, Switzerland

UNCTAD (1999) World investment report 1999, United Nations Conference on Trade and Development. Geneva, Switzerland

UNCTAD (2004) World investment report 2004, United Nations Conference on Trade and Development. Geneva, Switzerland

UNCTAD (2007) World investment report 2007, United Nations Conference on Trade and Development. Geneva, Switzerland

UNCTAD (2011) Global and regional FDI trends in 2010. Highlights, United Nations Conference on Trade and Development, issue no. 5. Geneva, Switzerland

UNCTAD (2012) World investment report 2012, United Nations Conference on Trade and Development. Geneva, Switzerland

Chapter 7
Concluding Remarks

7.1 Recapitulation

Apart from presenting the basic features characterizing the FDI phenomenon (in the initial chapter), its geographical and sectoral distribution and trends (in the previous one), the main aim of this work was that of discussing the issues associated to the relationship between FDI and the natural environment. For this reason, the three central chapters represent the core of our work and propose a review of the literature of the FDI-environment relationship. With reference to those research works and analysis activities produced within the context of the study of this relationship, it can be observed that literature is varied. In particular, it is composed of different perceptions and views which can be basically grouped into three main thematic areas: (1) the environmental effects of FDI flows; (2) the competition for FDI and its effects on environmental standards; (3) the cross-border environmental performance.

As has been reported in the dedicated chapter, the first area of discussion on the environmental effects of FDI comments on how FDI can generate benefits and costs for receiving countries. It is argued that FDI may boost economic growth, generate structural efficiency together with other positive effects but it is also recognized that FDI can also bring about environmental degradation. As a very brief reminder of what has been extensively said in the previous pages, it is thought that FDI brings new and more environmentally-friendly technologies to receiving countries. On the other hand, FDI is perceived as a contribution to the spread of industrial activity and as a driver of the production and consumption of industrial polluting goods. As has already been highlighted in the previous pages of this work, however, the practicability of an economic growth decoupled from environmental degradation responds to an unrealistic view. With regard to this, it must be stressed how, even in the case of the optimistic view considering the existence of a virtuous circle between FDI and the environment, FDI cannot be seen as a panacea and, particularly, as something capable of solving environmental problems. Growth—and induced-FDI growth as a consequence—always has a cost in terms of degradation of the natural environment. The consideration that the Earth-ecosystem is the base

P. Pazienza, *The Relationship Between FDI and the Natural Environment*,
SpringerBriefs in Economics, DOI: 10.1007/978-3-319-04301-2_7,

which human life—and, particularly, its economic activities—is based on and its use should be limited within certain thresholds, leads to one of the most important insights of ecological economics. In fact, it depends on the scale of the macro-economy relative to the ecosystem whether a growth path can be perceived economic or uneconomic.[1]

The second vein of discussion, namely the competition for FDI and its effect on environmental standards, whose literature has been analysed in the third chapter, basically considers two aspects: the impact of environmental standards on the location of a firm's investment decisions and the environmental effects of international countries' competition for FDI. Very briefly, the first aspect aims to verify if the existence of countries with different environmental regulations and standards can be a reason for firms relocating their activity. The second aspect analyses what happens to FDI when countries intentionally modify their environmental regulatory systems. As has already been said, this body of literature focuses on the following hypotheses: (1) "pollution havens"; (2) "race to the bottom"; (3) "regulatory chill". As has been clearly reported in the devoted chapter, the first hypothesis occurs when investors relocate their industries and activities to those countries where production is cheaper and more efficient in the light of environmental regulatory requirements. The literature shows little empirical evidence with regard to specific sectors. The second hypothesis consists, instead, of positive action by a government lowering environmental standards to bring in FDI. Empirical evidence is little and can be justified by the NIMBY syndrome. However, this hypothesis may only exist in certain sectors, such as that of the natural resources exploitation. Finally, the "regulatory chill" hypothesis occurs when the conviction that more stringent pollution control requirements impose costs that will arm the competitiveness of the regulated industry leads industrialists or politicians to argue for lower environmental standards (i.e. "the political drag" effect) than would be justifiable in the absence of competitiveness concerns.

The third and final considered cluster of research on cross-border environmental performance of foreign investing firms goes well beyond the understating of how a FDI location can affect or can be affected by environmental regulation. In fact, it focuses more on the identification of the approach of environmental management adopted by firms investing abroad, since this aspect is considered the most relevant in understanding the environmental implication of FDI in host economies. As a result, much of the research carried out within this framework basically focuses on investigating the driving forces behind the different environmental behaviours and performance of foreign firms in their cross-border and relatively undifferentiated operations. In other words, as has already been said in the specific chapter, it takes into account the aspect of corporate environmental performance while trying to understand the reason why a certain type of environmental management approach is chosen by a foreign investor once it has started its operations in a host country.

[1] According to Daly and Farley (2011), uneconomic growth occurs when further levels of growth of an economic system cost more than it is actually worth in environmental terms.

Having insofar recalled the core issues subject of attention in this work, it is now our aim to develop some considerations in order to highlight the gap of knowledge which still exists and calls for further research. The next section, which is the conclusive part of the work, will be dedicated to this purpose.

7.2 Ideas for a Research Agenda

With regard to the first thematic area on the environmental effects of FDI, we have observed how empirical research has found examples of both positive and negative marginal effects in some countries and not in others. This depends on specific circumstances associated to micro and macro aspects. In other terms, we are still unable to benefit from some final considerations which could help us definitively explain the FDI-environment relationship. This is particularly true for the identification of the "net effect" of FDI on the environment. As has already been said, this may be very difficult to do because of the complexity in discriminating the effects of the activity run by domestic industries from those of foreign firms or their affiliates and, also, because of the recognition that FDI does not occur as an isolated phenomenon. In fact, FDI does not only affect the environmental sphere, but it is also strictly linked to other factors. For this reason, as has already been noted in Chap. 3, most of the recent studies on the environmental effects of FDI are carried out by decomposing them into scale, composition (or structural) and technique effects. Apart from the fact that the literature contains more works focusing on the relationship between trade and investment rather than those investigating that of FDI and the environment, even this more recent analytical approach has not produced any final and definitive evidence. As a result, this investigation context is still and generally claimed to be one of the research areas characterized by the lack of better scientific understanding and calls for further research. To our personal observation, however, the inability of the literature to explain the considered relationship also depends on the fact that it fails to carry out empirical analysis with reference to specific sectors. In fact, as already highlighted, empirical investigation takes into consideration aggregated values of FDI flows and polluting agents disregarding the specific dynamic observable at the level of each specific activity sector. We do believe that more detailed analysis carried out while considering the breakdown of FDI and pollutants by activity sector would represent a concrete help in overcoming such a situation which in our view represents a heavy limitation to the development of more appropriate empirical analysis.

In relation to the second vein of discussion of competition for FDI and its effect on environmental standards, the analysis of the works available to us from the literature showed us how it represents that part of the literature where much of the debate on the FDI-environmental relationship has been developed. This is particularly true for the "pollution haven" hypothesis which discusses the location of a firm's investment decisions as a result of environmental regulation. As was highlighted in the devoted sections, however, environmental regulation is often seen to play a very minimal role in the decision process of investment location, although there is plenty of evidence that pollution-intensive industries do have location preferences for low environmental standards and are able to influence governments to create lenient environmental regulations. Other aspects such as size, growth and accessibility of potential markets, political stability, labour costs, easy repatriation of profits, red tape, education levels, infrastructures, administrative and legal transparency and certainty together with some other factors are found to play a more relevant role. However, as was adequately highlighted, empirical studies are still unable to reach a broadly accepted conclusion. As a result, the need for further research is still felt. In our view, this should try to use more detailed data, disaggregated at the sectoral level, instead of working on aggregated investment flow. This would help to drop the excessive focus on site-specific environmental impact and to go beyond the consideration of emissions related to a few industrial pollutants.

Finally, with regard to the third thematic area on cross-border environmental performance, we highlighted how the literature is unable to adequately explain why MNCs vary their environmental performance cross-country. Broadly speaking, from what can be read in the literature, this could be due to the lack of appropriate data and to the fact that the release of information on the operation of foreign firms abroad is very often a voluntary fact. It belongs to a general feeling in the literature that, as the search for public consent induces firms and investors to produce environmental reports of their activities, the problem of the gap of data will be covered. It is for this reason that this vein of discussion still results unexplored and a call for further research is generally felt.

References

Aden J, Kyu-hong A, Rock MT (1999) What is driving the pollution abatement expenditure behavior of manufacturing plants in Korea? World Dev 27(7):1203–1214

Antweiler W, Copeland BR, Taylor MS (2001) Is free trade good for the environment? Am Econ Rev 91:877–908

Baldwin RE (1994) The effects of trade and foreign direct investment on employment and relative wages. OECD Econ Stud 23:7–54

Bao Q, Chen Y, Song L (2011) Foreign direct investment and environmental pollution in China: a simultaneous equation estimation. Environ Dev Econ 16(1):71–92

Barbier E (1997) Introduction to environmental Kuznets curve special issue. Environ Dev Econ 2(4):369–381

Baumert KA, Herzog T, Pershing J (2005) Navigating the numbers: greenhouse gas data and international climate policy. World Resources Institute, Washington, D.C.

Ben Kheder S (2010) French FDI and pollution emissions: an empirical investigation. CERDI working papers, Clermont-Ferrand, France. Available at http://cerdi.org/uploads/sfCmsContent/html/323/BenKheder.pdf

Blackman A, Wu X (1999) Foreign direct investment in China's power sector: trends, benefits and barriers. Energy Policy 27(12):695–712

Blömstrom M, Kokko A (1996) Multinational corporations and spillovers, CEPR discussion paper no. 1365, London, U.K.
Boocock CN (2002) Environmental impacts of foreign direct investment in the mining sector in sub-saharan Africa. In: Foreign direct investment and the environment: lessons from the mining sector. OECD, Paris, France
Burns C, Potts M (2011) The "rebound effect": a perennial controversy rises again. eSolutions J 4(2). Available at http://www.rmi.org/
Cantwell J, Bellack C (1998) How important is foreign direct investment? Oxford Bull Econ Stat 60:99–106
Chaisrisawatsuck S, Chaisrisawatsuck W (2007) Imports, exports and foreign direct investment interactions and their effects. Asia-Pacific research and training network on trade. Working paper series no. 45. Available at http://www.unescap.org/tid/artnet/pub/wp4507.pdf
Chen JR, Yang CH (1999) Determinants of foreign direct investment: comparison between expansionary FDI and defensive FDI. Taiwan Econ Rev 27:215–240
Chen TJ, Ku YH (2000) The effect of foreign direct investment on firm growth: the case of Taiwan's manufacturers. Jpn World Econ 12:153–172
Christmann P, Taylor G (2001) Globalization and the environment: determinants of self-regulation in China. J Int Bus Stud 32(3):439–458
Coe DT, Helpman E (1995) International R&D spillovers. Eur Econ Rev 39:859–887
Cole MA, Elliot RJ (2003) Determining the trade-environment composition effect: the role of capital, labor and environmental regulations. J Environ Econ Manag 46(3):363–383
Cole MA, Elliot RJ (2005) FDI and the capital intensity of 'dirty' sectors: a missing piece of the pollution Haven puzzle. Rev Dev Econ 9(4):530–548
Cole MA, Elliot RJ, Fredriksson PG (2006) Endogenous pollution Havens: does FDI influence environmental regulation? Scand J Econ 108(1):157–178
Common M, Perrings C (1992) Towards an ecological economics of sustainability. Ecol Econ 6:7–34
Daly HE, Farley J (2011) Ecological economics: principles and applications, 2nd edn. Island Press, Washington, D.C.
Dasgupta S, Hettige H (2000) What improves environmental performances. Evidence from the Mexican industry. J Environ Econ Manag 39(1):39–67
Dasgupta S, Mody A, Roy S, Wheeler D (2001) Environmental regulation and development: a cross-country empirical analysis. Oxford Dev Stud 29(2):173–187
Dessus S, Bussolo M (1996) Politiques de l'environment et libéralisation des échanges au Costa Rica: une vue d'ensemble. OECD Development Center. Technical working paper no. 110, Paris, France
Dorsey T (2008) A capital story. IMF Finan Dev 45(2):46–49
Elliot RJR, Shimamoto K (2008) Are ASEAN countries Havens for Japanese pollution-intensive industry? World Econ 31(2):236–254
Eskeland GS, Harrison AE (2003) Moving to greener Pastures? Multinationals and the pollution-Haven hypothesis. J Dev Econ 70(1):1–24
Esty DC, Gentry BS (1997) Foreign direct investment, globalization and environment. In: OECD, globalization and environment: preliminary perspectives. OECD, Paris, France
Esty DC, Geradin D (1998) Environmental protection and international competitiveness. J World Trade 32(3):5–47
French H (1998) Investing in the future: harnessing private capital flows for environmentally sustainable development. Worldwatch paper no. 139. Worldwatch Institute, Washington, D.C., USA
Gassebner M, Lamla MJ, Sturm J (2011) Determinants of pollution: what do we really know? Oxford Econ Papers 63:568–595
Gentry B (ed) (1998) Private capital flows and the environment. Lessons from Latin America. Edward Elgar Publishing, Cheltenham
Gentry B (1999) Foreign direct investment: boon or bane? In: Proceedings of OECD, foreign direct investment and the environment. OECD, Paris

Gray KR (2002) Foreign direct investment and environmental impacts: is the debate over? Rev Eur Commun Int Environ Law 11(3):306–313

Greening LA, Greene DL, Difiglio C (2000) Energy efficiency and consumption—the rebound effect—a survey. Energy Policy 28(6–7):389–401

Grossman GM, Krueger AB (1991) Environmental impacts of the North American free trade agreement. NBER working paper no. 3914

Grossman GM, Krueger AB (1993a) Environmental impacts of the North American free trade agreement. In: Garber PM (ed) The Mexico-US free trade agreement. MIT Press, Cambridge

Grossman GM, Krueger AB (1993b) Pollution and growth: what do we know? In: Goldin I, Winters L (eds) The economics of sustainable development. MIT Press, Cambridge

Grossman GM, Krueger AB (1995) Economic growth and the environment. Q J Econ 110(2):353–377

Guoming X, Cheng Z, Yangui Z, Shunqi G, Zhan JX (1999) The interface between foreign direct investment and the environment: the case of China. Report as part of the UNCTAD/DICM project. Cross border environmental management in transnational economies. Occasional paper no. 3. Copenhagen Business School

Hansen MW (1999) Cross border environmental management in transnational corporations. An analytical framework. Report as part of the UNCTAD/CBS project. Cross border environmental management in transnational economies, Occasional paper no. 5. Copenhagen Business School

He J (2006) Pollution Haven hypothesis and environmental impact of foreign direct investment: the case of industrial emission of sulfur dioxide (SO_2) in Chinese provinces. Ecol Econ 60:228–245

Hejazi W, Safarian E (2001) The complementarity between U.S. foreign direct investment stock and trade. Atlantic Econ J 29(4):420–437

Hettige H, Huq M, Pargal S, Wheeler D (1996) Determinants of pollution abatement in developing countries: evidence from South and Southeast Asia. World Dev 24(12):1891–1904

Huq M, Wheeler D (1993) Pollution reduction without formal regulation: evidence from Bangladesh, World Bank. Working paper no 39. World Bank, Washington, D.C.

Ignacio M (2003) Foreign direct investment and the protection of the environment: the border between national environmental regulation and expropriation. Eur Environ Law Rev 12(3):70–82

IMF (1993) Balance of payments manual. International Monetary Fund, Washington, D.C.

Jackson T (2009) Prosperity without growth. Economics for a finite planet. Earthscan, London

Jha V (1999) Investment liberalization and environmental protection: conflicts and compatibilities in the case of India. Report as part of the UNCTAD/DICM project. Cross border environmental management in transnational economies. Occasional paper, Copenhagen Business School

Liang FH (2006) Does foreign direct investment harm the host country's environment? Evidence from China. Working paper of the Haas School of Business, University of California, Berleley, U.S.A.

Mabey N, McNally R (1998) Foreign direct investment and the environment. From pollution havens to sustainable development. WWF-UK Report

Meers R (2000) A test of the environmental Kuznets curve for local and global pollutants, Honors projects, paper no. 76 of the Economic Department of the Illinois Wesleyan University

Mihci H, Cagatay S, Koska O (2005) The impact of environmental stringency of the foreign direct investments of the OECD countries. J Environ Assess Policy Manag 7(4):679–704

Mihci H, Cagatay S, Koska O (2006) FDI-trade-environment linkages: changing flows between EU(15), enlarged EU and Non-European countries, discussion paper for the 8th European treade study group (ETSG) annual conference. Vienna, Austria

Moosa IA (2002) Foreign direct investment: theory, evidence and practice. Palgrave MacMillan Ltd., New York

Motta M, Thisse JF (1994) Does environmental dumping lead to delocation? Eur Econ Rev 38:563–576

Munasinghe M (1999) Is environmental degradation an inevitable consequence of economic growth? tunnelling through the environmental Kuznets curve. Ecol Econ 29:89–109

Murray CK (2013) What if consumers decide to all "go green"? environmental rebound effects from consumption decisions. Energy Policy 54:240–256

Newell P (2001) Managing multinationals: the governance of investment for the environment. J Int Dev 13:907–919

Nguyen CD, Simpson G, Saal D, Nguyen AN, Pham NQ (2008) FDI horizontal and vertical effects on local firm technical efficiency, DEPOCEN working paper no. 17, pp 1–22

Nordström H, Vaughan S (1999) Trade and environment, world trade organization, special studies no. 4. Geneva, Switzerland

O'Connor D (2000) Global capital flows and the environment in the 21st century, organization for the economic co-operation and development, technical paper no. 161. Paris, France

OECD (1997) Foreign direct investment and the environment: an overview of the literature, document of the negotiating group on the multilateral agreement on investment (MAI). OECD, Paris, France

OECD (2001) Environmental outlook, organization for economic co-operation and development. Paris, France

OECD (2002) Environmental issues in policy-based competition for investment: a literature review, document of the working party on global and structural policy. OECD, Paris, France

OECD (2002b) Environmental benefits of foreign direct investment: a literature review, document of the working party on global and structural policy. OECD, Paris, France

OECD (2007) Glossary of statistical terms. OECD, Paris, France

Opschoor JB (1995) Ecospace and the fall and rise of throughput intensity. Ecol Econ 15:137–140

Panayotou T (1997) Demystifying the environmental Kuznets curve: turning a black box into a policy tool. Environ Dev Econ 2(4):465–484

Panayotou T (2000) Globalization and the environment, center for international development working paper no. 53. Harvard University, Boston

Perman R, Stern DI (2003) Evidence from panel unit root and cointegration tests that the environmental Kuznets curve does not exist. Aust J Agric Res Econ 47(3):325–347

Porter G (1999) Trade competition and pollution standards: 'race to the bottom' or 'stuck at the bottom'? J Environ Dev 8(2):133–151

Porter ME (1990) The competitive advantage of nations. Free Press, New York

Rasiah R (1999) Transnational corporations and the environment: the case of Malaysia, report as part of the UNCTAD/CBS project cross border environmental management in transnational economies, occasional paper no. 4. Copenhagen Business School

Repetto R (1995) Jobs, competitiveness and environmental regulation: what are the real issues?. World Resources Institute, Washington, D.C

Revesz RL (1992) Rehabilitating interstate competition: rethinking the 'race to the bottom' rationale for federal environmental regulation. NY Univ Law Rev 67:1210–1254

Selden TM, Song D (1994) Environmental quality and development: is there a Kuznets curve for air pollution emissions? J Environ Econ Manag 27:147–162

Shafik N (1994) Economic development and environmental quality. Oxford Econ Pap 46(Supplementary):757–773

Shafik N, Bandyopadhyay S (1992) Economic growth and environmental quality. Time series and cross-country evidence, policy research working paper no. 904. World Bank, Washington

Shahbaz M, Nasreen S, Afza T (2011) Environmental consequences of economic growth and foreign direct investment: evidence from panel data analysis, MPRA paper no. 32547

Sinani E, Meyer KE (2004) Spillovers of technology transfer from FDI: the case of Estonia. J Comp Econ 32(3):445–466

Spatareanu M (2007) Searching for pollution havens: the impact of environmental regulations on foreign direct investment. J Environ Dev 16(2):161–182

Stern DI (2004a) Environmental Kuznets curve. In: Cleveland CJ (ed) Encyclopedia of energy. Elsevier Inc., Amsterdam

Stern DI (2004b) The rise and fall of the environmental Kuznets curve. World Dev 32(8):1419–1439
UN (2011) World economic situations and prospects 2011. United Nations Publication, New York
UN (2012) World economic situations and prospects 2012. United Nations Publication, New York
UNCTAD (1993) World investment report 1993. In: United Nations conference on trade and development. Geneva, Switzerland
UNCTAD (1999) World investment report 1999. In: United Nations conference on trade and development. Geneva, Switzerland
UNCTAD (2004) World investment report 2004. In: United Nations conference on trade and development. Geneva, Switzerland
UNCTAD (2007) World investment report 2007. United Nations conference on trade and development. Geneva, Switzerland
UNCTAD (2011) Global and regional FDI trends in 2010. In: Highlights, United Nations conference on trade and development, Issue no. 5. Geneva, Switzerland
UNCTAD (2012) World investment report 2012. United Nations conference on trade and development. Geneva, Switzerland
UNEP (2007) Global environment outlook GEO4: environment for development. United Nations Environment Programme, Nairobi, Kenya
Vogel D (1995) Trading up: consumer and environmental regulation in a global economy. Harvard University Press, Cambridge
Wenhua D (2007) Pollution abatement cost savings and FDI inflows to polluting sectors in China. Environ Dev Econ 12(6):775–798
Wheeler D (2001) Racing to the bottom? Foreign investment and air pollution in developing countries. J Environ Dev 10(3):225–246
World Bank (2000) Greening industry. New roles for communities, markets and governments. World Bank and Oxford University Press, Washington, D.C.
WTO (1998) World trade organisation annual report. World Trade Organization, Geneva, Switzerland
WWF (2001) Preliminary assessment of the environmental and social effects of liberalization of tourism services. WWF international discussion paper, May, Geneva, Switzerland
Xing Y, Kolstad CD (2002) Do lax environmental regulations attract foreign investment? Environ Res Econ 21:1–22
Yandle B, Vijayaraghavan M, Bhattarai M (2002) The environmental Kuznets curve: a primer, property and environment research centre (PERC) study no. 02–01. Available at http://www.perc.org/articles/article688.php
Zarsky L (1999) Havens, halos and spaghetti: untangling the evidence about foreign direct investment and the environment. In: OECD, foreign direct investment and the environment, conference proceedings. OECD, Paris

Visited Web-Pages:

http://unctad.org/en/Pages/DIAE/FDI%20Statistics/SOurces-and-definitions.aspx

MIX
Papier aus verantwortungsvollen Quellen
Paper from responsible sources
FSC® C105338

If you have any concerns about our products, you can contact us on
ProductSafety@springernature.com

In case Publisher is established outside the EU, the EU authorized representative is:
Springer Nature Customer Service Center GmbH
Europaplatz 3, 69115 Heidelberg, Germany

Printed by Libri Plureos GmbH
in Hamburg, Germany